An Applied Mathematician's Apology

An Applied Mathematician's Apology

Lloyd N. Trefethen
University of Oxford

Society for Industrial and Applied Mathematics
Philadelphia

Publications Director	Kivmars H. Bowling
Executive Editor	Elizabeth Greenspan
Acquisitions Editor	Elizabeth Greenspan
Developmental Editor	Mellisa Pascale
Managing Editor	Kelly Thomas
Production Editor	Kelly Thomas
Production Manager	Donna Witzleben
Production Coordinator	Cally A. Shrader
Graphic Designer	Doug Smock

Library of Congress Cataloging-in-Publication Data
CIP data is available at *www.siam.org/books/ot182*

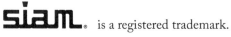 is a registered trademark.

In memory of Carl Runge, a kindred spirit from another era.

Contents

A Note about the Title

G. H. Hardy published *A Mathematician's Apology* in 1940. The sense of the term is that of apologia, a defense of a field. It could be said (and some of my friends have said) that a more accurate title for the present piece would have been *Confessions of a Numerical Analyst*. To be sure, this essay differs in many ways from Hardy's, containing more biographical material and also more mathematics, especially in the second half. But its purpose is the same, a serious and personal meditation about mathematics.

1. Introduction

I am a passionate mathematician, but I am puzzled. My life is mathematics, and I feel a strong connection to the mathematicians of the past. As the years go by and I work on new problems and gain in knowledge and perspective, my sense of myself as a mathematician just gets stronger. And yet I feel a disconnection from the mathematics and the mathematicians of the present.

I began to write this essay as a means to explore this odd situation. Naturally enough, I began by reflecting on the experiences of my career, and before long I found I was writing a memoir, too. It is the story of a mathematician in an unusual (and very lively) corner of the subject.

My part of mathematics is numerical analysis, which I defined in an essay thirty years ago like this:

Numerical analysis is the study of algorithms for the problems of continuous mathematics.

(Continuous means involving real or complex numbers.) The traditional idea is that other mathematicians might invent the notion of the roots of a polynomial, say, and then it is up to the numerical analysts to develop algorithms to calculate them. For example, the polynomial $x^5 + x^3 - 3$ is equal to 0 if $x = 1.105298546006169\ldots$. How do we calculate those digits? By executing algorithms developed by numerical analysts. Of course, mathematicians have invented many more complicated problems than roots of polynomials, such as partial differential equations, which are the basis of much of the natural sciences.

Numerical analysts are tasked with solving these too, and scientists and engineers use our methods all the time.

Newton, Euler, and Gauss were outstanding numerical analysts back in an era when it was self-evident that part of the business of mathematicians was to calculate things. But the landscape has changed since then, with other branches of mathematics appearing and flourishing to a degree unimaginable in those days. Nowadays, most leading mathematicians have little interest in calculation, which they avoid by habit and may disdain as unimportant in principle. They work on other things, and no paper by a researcher like me would appear in a top journal like *Annals of Mathematics*. Meanwhile numerical analysis thrives separately, and of course, we have plenty of journals of our own. Demographically we are big, accounting for perhaps 5% of academic mathematicians, and in impact on science and technology we are enormous.

My personal good fortune has been remarkable. I hold what is arguably the most visible chair in my field in the world, the Professorship of Numerical Analysis at the University of Oxford. This big mathematics department lists 100 professors on the web site and is generally rated in the top group along with Harvard, MIT, Stanford, Berkeley, Cambridge, and Princeton. None of those other universities has a chair in Numerical Analysis, but Oxford does, and since 1997, the Professor of Numerical Analysis has been me. Our Numerical Analysis Group has been a leader in the subject in Britain since its founding in the 1960s and is well known around the world. I personally am well known, too, author of widely read textbooks and technical papers, Fellow of the Royal Society, former President of the Society for Industrial and Applied Mathematics (SIAM), winner of big prizes and honorary degrees. I am a fellow of Balliol College, founded in the days of Kublai Khan.

Obviously this is a success story, and indeed, it could hardly be better. It doesn't sound like the profile of one who feels disconnected from his discipline. So what is going on?

2. Mathematics in Childhood and High School

The love of the subject begins when you're a kid. I grew up in Lexington, Massachusetts, and like most future mathematicians, I found it easy to get the right answers in school—to do "sums", as we say in England, though as an American I still find that expression foreign. I remember exercise sheets with empty boxes in them, like $5 + \square = 12$, and you had to figure out the missing number. That was easy, and it was funny that some of my classmates had trouble with it. Most mathematicians have memories like these.

Traveling around the world with my parents and sister at age 9 for my father's sabbatical year 1964-5, I missed fourth grade at Shady Hill School, but after 28 days crossing the Pacific on a freighter with just 11 passengers, I enrolled at Seaforth Primary School in Sydney, Australia, where again the math came easy. To keep us at the level of our friends back home, it was arranged that my mother would teach me and Gwyned extra English and my father would teach us extra math. This wasn't hard for him, since he was a professor of mechanical engineering at Tufts University, and in fact, during these months in Sydney, he led a team that for the first time in the Southern Hemisphere drained a "bathtub" under carefully enough controlled conditions to observe the Coriolis effect. In my recollection, mathematics lessons with my father amounted to leisurely afternoons at his side in the lounge of the passenger ship *Ellinis* we took from Brisbane to Athens in May 1965. (We passed through the Suez Canal, two years before it was closed in the 1967 war.) The subject of our study was negative numbers, which were explained with the help of bugs on a number line. For example, suppose a bug is at position -5, facing to the left, and it hops backwards three units. Where does it end up? At -2, of course. This explains why $-5 - (-3) = -2$, and I found this kind of thing easy and fun. After returning to Shady Hill at age 10, I had the odd sense that simply knowing how to add, subtract, multiply, and divide negative numbers put me about three years ahead of the rest of the class.

Not ahead of Nat Foote, however, the red-headed boy who had joined the school during my absence. He and I were the math whizzes in the Shady Hill Class of 1970, and in 7th, 8th, and 9th grades, the two of us were taken out of the regular classroom to study independently from textbooks under the guidance of the teacher, Bob Lawler. We learned a lot of algebra. My father had never taught me that amazing technique called factoring, like $x^2 - 2x - 3 = (x + 1)(x - 3)$, whereas Nat had learned it from his brother George. We studied trigonometry, too, so I was good at sines and cosines. Nat and I tended to be boisterous when left unsupervised for our math hour, and I remember arguing with Mr. Lawler about mathematical things. He told me that 1 divided by 0 was "undefined," and I said that was stupid, obviously it was infinity.

At age 15 Nat and I went to an outstanding high school, Phillips Exeter Academy, which even in the 1970s had three PhDs on the math faculty. In our first class on the first day, which was a calculus course filled mostly with seniors, Mr. Lynch taught us the definition of the derivative of a function as a limit in the style of epsilons and deltas, which he wrote carefully on the blackboard. Wow! I had never seen this before, though Nat had learned it from George. The derivative seemed truly serious, requiring concentrated thought, and I remember thinking that if big ideas were going to come at this rate in high school, it was going to be a pretty intense experience. Nat and I went on to pass the BC calculus advanced placement test that spring.

I became hooked on computers that year, for Exeter had teletypes connected to the Dartmouth Time-Sharing System. At first I figured that everybody was using them, and there was no need for me to join the crowd; but a few weeks later I'd given it a try, and that was that. Mathematics was the obvious application to explore, and I remember writing a sequence of BASIC programs to print prime numbers 2,3,5,7,…, each more efficient than the last. There were not enough terminals to go around, and I often skipped lunch to grab an available space, but I think there was only one day when I skipped both lunch and dinner.

Then came another sabbatical year travelling around the world with my parents. This experience was enriched by a special contribution from Steve Maurer, a teacher who was also a PhD student at Princeton, who had become a friend through having breakfast most mornings with me and Nat that first year at Exeter in the Elm St. Dining Hall. Mr. Maurer put together a canary sheaf of 33 difficult "Problems for a World Tour" for me to work on as I travelled, and these became a theme of my junior year at large. However, I managed only five or six of the problems, and I felt this was a sign of inadequacy. At Mr. Maurer's suggestion I also studied the early chapters of Feller's classic text on probability, exciting material, and when our family paused for four months in Seattle en route to revisiting Australia, I enrolled in a very mechanical linear algebra course at the University of Washington and also an honors analysis course from an inspiring professor, Carl Allendoerfer. My contemporary Bill Gates was living a mile or so from us at the time, attending Lakeside School and doing equally advanced math along with his other activities, but I hadn't heard of him yet.

Back for senior year at Exeter, Nat and I were able to fly high. In the first semester we took abstract algebra from David Arnold using Fraleigh's textbook, and this was the most thrilling mathematical experience I had ever had. The definition of a group was so beautiful! Mr. Arnold let Nat and me do a special project on the Sylow Theorems. In our final semester we were then introduced to a subject which was to prove important to my career. We got to choose a "field course" on whatever topic we wanted, and we picked complex analysis, that is, the mathematics of real and imaginary numbers and the functions built from them. Our instructor was the very special David Robbins, who was teaching for a few years before beginning his career at the Institute for Defense Analysis, and the textbook was the classic by Churchill. In this course I remember having an edge over Nat, which felt good. Mr. Robbins put a problem on one of the tests involving an ant that moves one unit, then turns 30 degrees left and moves half a unit, then turns another 30 degrees left and moves a quarter unit, and so on; where does it end up? (Bugs again, eight years later!—but now in the complex plane and with a full compass of motion, not just forward or reverse.) I remember being pleased that I spotted that this was a power series and Nat didn't. On the whole, though, Nat and I were more or less indistinguishable all-rounders. He graduated first in the class, and I was second. I got the math prize. Though Mr. Maurer advised against it (one must make new friends), we decided to continue as roommates at Harvard.

At age 18, there was little doubt I was heading to be a mathematician. I remember that summer, before Harvard, driving to a park a few miles from home in Lexington with Herstein's abstract algebra book so I could study that subject in depth. This was not successful, and I got sleepy in the hot sun.

Many mathematicians have stories like these of early years. Most of us found ourselves good at the subject without trying very hard, helped along by a special teacher or two, and by one means or another ended up learning things beyond the usual curriculum. In 1973 I placed first in the New Hampshire High School Mathematics test, getting the first-ever perfect score, and Nat came second. But when this led to the US Mathematical Olympiad for high school students, I didn't do so well. All in all, I was very good, but never spectacular. Nor did I have any of the out-of-school mathematical training that many kids were already benefiting from and which later became an industry, like summer math camps and coaching sessions for competitions. I just pursued the subject eagerly, encouraged by supportive parents, outstanding teachers, and a best friend and rival who was every inch my equal.

Here is something funny from that senior year at Exeter. I had jumped at the opportunity to go to such a special school, never seriously considering the option of staying home in Lexington. Well, on the National High School Mathematics Test that year, whereas Exeter ranked #2 in New England, Lexington High School was #1! We joked that if I had stayed home, it might have been the other way around.

At age 18, I don't think I questioned any aspect of the subject of mathematics. Math was there to be studied, and I was a student. The experts had developed it over the ages, and it was my lucky opportunity to learn some of what they had discovered. Nor did I have an idea that a debate about "pure and applied" would one day be important to me.

3. Fields Medalists and Their Strangely Small Impact on Me

Forward half a century. Here is a list of all sixty Fields medalists so far (including Grigori Perelman in 2006, though he declined the award). These are the gods of mathematics. All are brilliant. Some have an aura of being more than that, gods among the gods.

1936	Lars Ahlfors	1936	Jesse Douglas
1950	Laurent Schwartz	1950	Atle Selberg
1954	Kunihiko Kodaira	1954	Jean-Pierre Serre
1958	Klaus Roth	1958	René Thom
1962	Lars Hörmander	1962	John Milnor
1966	Michael Atiyah	1966	Paul Cohen
1966	Alexander Grothendieck	1966	Stephen Smale
1970	Alan Baker	1970	Heisuke Hironaka
1970	Sergei Novikov	1970	John Thompson
1974	Enrico Bombieri	1974	David Mumford
1978	Pierre Deligne	1978	Charles Fefferman
1978	Grigori Margulis	1978	Daniel Quillen
1982	Alain Connes	1982	William Thurston
1982	Shing-Tung Yau	1986	Simon Donaldson
1986	Gerd Faltings	1986	Michael Freedman
1990	Vladimir Drinfeld	1990	Vaughan Jones
1990	Shigefumi Mori	1990	Edward Witten
1994	Jean Bourgain	1994	Pierre-Louis Lions
1994	Jean-Christophe Yoccoz	1994	Efim Zelmanov
1998	Richard Borcherds	1998	Timothy Gowers
1998	Maxim Kontsevich	1998	Curtis McMullen
2002	Laurent Lafforgue	2002	Vladimir Voevodsky
2006	Andrei Okounkov	2006	Grigori Perelman

2006	Terence Tao		2006	Wendelin Werner
2010	Elon Lindenstrauss		2010	Ngô Bào Châu
2010	Stanislav Smirnov		2010	Cédric Villani
2014	Artur Avila		2014	Manjul Bhargava
2014	Martin Hairer		2014	Maryam Mirzakhani
2018	Caucher Birkar		2018	Alessio Figalli
2018	Peter Scholze		2018	Akshay Venkatesh

Now I'm no god, but remember, I am a leading figure in one of mathematics' big subdisciplines and the senior among 15 statutory chairs in one of the world's top mathematics departments. So let's ask, how many works by these Fields medalists have I read?

The answer is, exactly one. This is the inspiring textbook *Complex Analysis* by Ahlfors that I studied in Math 213a, sophomore year at Harvard. Beyond that, I've read some pages by Serre, Hörmander, Milnor, Atiyah, Smale, Lions, and Tao. But Ahlfors, Fields medalist 86 years ago, is the only one of the 60 who wrote something I have come anywhere close to reading in full.

All areas of intellectual activity have proliferated in specializations with the years, as more and more people make contributions, but still, this situation is extreme. Can you imagine a novelist who's never read a book by Mann, Hemingway, Márquez, Lessing, or Morrison? An economist who's never read anything by Samuelson, Arrow, Friedman, Kahneman, or Krugman?

Part of the effect is related to the pure vs. applied divide, which I will keep coming back to. I am applied, and these Fields medalists are pure. Officially, Fields medals are for "mathematics," but it is understood that applied math doesn't count, even though you'll never find such a statement in print. Many of the winners would probably claim that they are simply mathematicians, that the distinction between pure and applied is illusory, and I'll have a word to say about that view later on. The Fields medalist with the clearest record of applied contributions is probably David Mumford, but these came after his Fields medal, beginning at age 45 essentially as a second career at a second university.

But pure vs. applied alone isn't enough to explain the weakness of the link between numerical people like me and the anointed leaders of mathematics. We might think of an applied mathematician as analogous to an experimental rather than theoretical physicist. Can you imagine an experimental physicist who's never read a paper by Einstein, Schrödinger, Bethe, Feynman, or Glashow? (I decided to get some data on this question, so I asked a few friends in experimental physics about their reading histories. It seems it is typical to have read 5–10 works by these five men.)

It's not that the Fields medalists have zero impact on me. I've talked with Ahlfors (he was a reader of my undergraduate thesis), Smale, Mumford (he taught me Math 250b at Harvard), Cohen (one of my professors at Stanford), Gowers, Tao, Smirnov, and Hairer. I've attended lectures by Atiyah, Fefferman, Thurston, Yau, Donaldson, Witten, Lions, and Villani. I have a distant impression of one or two of the contributions of Thompson, Werner, Perelman, and Mirzakhani. But as we say in mathematics, the impact of these people on me has been epsilon. Whoever they are influencing so greatly as to deserve Fields medals, it is not the Professor of Numerical Analysis at Oxford. And, of course, they have been equally little influenced by me. It would be interesting to know how many of the sixty have read a work of mine, and in the absence of data on this point, I would estimate this number as approximately 1.

Some Fields medalists live the glamorous academic life, giving distinguished lectures all over the place, but one who was not like that was the reclusive Dan Quillen, with whom I have a curious connection. I was on a mathematics faculty with Quillen not once but twice—at MIT and then again later at Oxford before his retirement in 2006—and I never met him.

4. Undergraduate at Harvard: Choice of Numerical Analysis

At Harvard, Nat and I entered as two of the freshman hotshots in math. The ethos was: advanced courses, advanced courses! Nobody serious about mathematics would ever sink so low, for example, as Math 21, linear algebra. So I first heard the linear algebra word "eigenvalue" in a physics course. Eigenvalues would play an important role in my career.

Most of the hotshots, including Bill Gates, took the legendary Math 55, taught by John Mather, which was designed to exhibit one's testosterone.[1] After long discussion of pros and cons, however, Nat and I decided to deviate from this path and take Math 105, a third-year analysis course taught by Neil Fenichel. For this we studied word-by-word, with intense effort and excitement, the opening chapters of Dieudonné's *Foundations of Modern Analysis*. Following the Bourbaki tradition, and living up to his surname, Dieudonné announces at the beginning that there will be no figures in the book, since figures encourage unrigorous thought. Here is how he puts it:

> This has also as a consequence the necessity of a strict adherence to axiomatic methods, with no appeal whatsoever to "geometric intuition," at least in the formal proofs: a necessity which we have emphasized by deliberately abstaining from introducing any diagram in the book. My opinion is that the graduate student of today must, as soon as possible, get a thorough training in this abstract and axiomatic way of thinking if he is ever to understand what is currently going on in mathematical research.

This extreme Bourbaki point of view, I hasten to add, has lost favor in later decades even among pure mathematicians.

I still have Dieudonné on my shelf, every line of the early chapters highlighted in orange, yellow, and blue to distinguish definitions, theorems, and other important material. It was a big course, with 40 or more students, and Nat and I got two of the four A+ grades. Meanwhile we were also taking Physics 55 along

---

[1] See "Math 55" at Wikipedia—and footnote 8.

with Steve Ballmer and Jim Sethna, among others. Every semester at Harvard, I enrolled in both math and physics courses, and what I learned in those courses laid the foundation of my career. As a rule I got A's in math courses and A - 's in physics, and I felt this pattern reflected reasonably my talents.

A big bifurcation came at the end of freshman year, when Nat announced that he was moving into the social sciences. Up to that point, we had been taking almost exactly the same courses since age 10. After 1974 we diverged, he following his father into the business world, me following mine into science and academics. Like most Harvard graduates, Nat has gone on to earn vastly more money than I have in his career, but money has only occasionally been a constraint for me, and although I note the gap, it doesn't disturb me too much.

In sophomore year the serious young mathematicians, now without Nat, hurried into graduate courses. I took Math 213a/b, complex analysis, with Barry Mazur and Raoul Bott, and Math 250a/b, abstract algebra, with John Tate and David Mumford. Though I didn't fully appreciate it then, I later came to see that learning mathematics from this incandescent foursome was about like being taught rock and roll by John, Paul, George, and Ringo. I liked these professors very much—what an intense year! But it was hard going, and these classes had plenty of graduate students in them who knew more than I did, including Putnam Exam winners. I recognized that two of the other undergraduates were better mathematicians than I was: Tom Goodwillie, a year older than me and now a professor at Brown University, and Nat Kuhn, son of the philosopher of science Thomas Kuhn, a year younger and now a psychiatrist in the Boston area.

Math 250a, with Tate, was the only course I took with Bill Gates. The lectures were late mornings Tuesdays and Thursdays, and afterwards a bunch of us would go to lunch, typically a subset of me, Bill, Doug Critchlow (now at Ohio State), Nat Kuhn, Tom Goodwillie, and somebody called Jack whose last name I don't recall. Sometimes we ate at Currier House, where Bill lived. He was thin, quick, and confident. At one of these lunches, Bill bragged that he was the fastest typist around since he used a keyboard all day long. I firmly disagreed, so we made a bet about it and trooped up to his room with two witnesses to see who was really the fastest. My touch-typing smashed his hunt-and-peck.

Another memorable person in my class at Harvard was Paul Ginsparg, an impressive (and even more confident) young physicist. Ginsparg went on to found arXiv, the e-Print repository that launched the era of open-access publishing.

At the end of this intense sophomore year came my personal bifurcation. All along I had assumed, as young mathematicians do, that I would pursue pure mathematics. After all, this was the very core of the subject, the place to make one's mark for all time. I remember thinking that whereas some people concentrated on just analytic number theory or algebraic number theory, I was more serious than that and would be a star of both analytic *and* algebraic number theory.

Somehow by the end of sophomore year, all this fell away, and I decided that what mattered was applied mathematics. These many years later, I find it surprisingly hard to recall details of how my views changed. I think I had a sense that applied mathematics was simply more connected with the world than pure, and without a doubt, the influence of my father played a part. I don't mean that he tried to persuade me one way or another, but it was he who had formed me as a thinker with discussions of scientific questions throughout my childhood and teenage years. From him, I had always had a sense of mathematics as one point on a spectrum including physics, chemistry, and engineering. (I don't think I took biology seriously at that age.)

So I switched majors from Math to Applied Math and arranged to start taking applied courses in the fall. One of these was Applied Math 211, the graduate course in numerical analysis, energetically taught by Don Rose from the marvelous books by Forsythe & Moler and Dahlquist & Björck (three of whom I would come to know pretty well). I was well ready for this, since I'd been programming Fortran for Howard Emmons' Home Fire Project for several years, ten hours a week during university terms and full-time during the summer. Other than Gates, incidentally, there were probably only three or four members of the Harvard Class of `77 with as much programming experience as I had.

This was it, I discovered. Numerical analysis was the heart of applied mathematics, the heart of mathematics, the heart of science. I remember the thrill of a late night with the PDP-8 in the Engineering Sciences Lab in which I finally got my Fortran code working to solve a boundary-value problem by the shooting method. It converged, and the correct digits lined up proudly. In my daily index card of November 4, 1975, at age 20 plus a couple of months, I wrote: "AM 211. I love this course."

5. The Field's Odd Reputation

Numerical analysis, the field I would devote my life to. A subject that may appear to be just one of mathematics' many subdisciplines, one of 16 if you go by the list of research groups at our departmental website. And yet, forty years of working in this area have given me a special vision of mathematics. We numerical people are the ones who see the show live. It happens on our screens and at our fingertips. We *make it happen*. The energy of this experience has kept me going, decade after decade, and it's always something of a mystery to me why more mathematicians don't recognize numerical computation as an indispensable way to explore mathematics.

A friend who knew me well asked after my first few months of graduate school, "Are you skiing on the complex functions yet?"

But I want to comment here on a public relations problem of this subject of mine. These observations come from my 1992 essay mentioned earlier, "The definition of numerical analysis," and I spoke about them in my inaugural lecture at Oxford in 1998.

When computers deal with real numbers like π and $\sqrt{2}$, they usually approximate them to about 16 digits of accuracy. The approximations entail what are called rounding errors, which occur all the time, whenever you calculate anything numerically.

Rounding errors have their interest, but they are pretty ugly from the usual mathematical point of view, and somehow or other, this ugliness came to be regarded as the very essence of my field. In writing that essay I looked up dictionary definitions of numerical analysis and recorded these dispiriting specimens:

> Webster's New Collegiate Dictionary (1973): *The study of quantitative approximations to the solutions of mathematical problems including consideration of the errors and bounds to the errors involved.*

> Chambers 20th Century Dictionary (1983): *The study of methods of approximation and their accuracy, etc.*

> The American Heritage Dictionary (1992): *The study of approximate solutions to mathematical problems, taking into account the extent of possible errors.*

How dreary! In the inaugural lecture, I had some fun imagining what it would be like if other fields had our flair for publicity:

> Aeronautical engineering. *The design of flying vehicles capable of rapidly and reliably reducing the errors in passengers' initial positions.*

> Education. *The development of methods for measuring the ignorance of children and adults, and diminishing it where possible.*

> Medicine. *The study of the approach of death in humans, and of methods for delaying this event.*

I think I know how the sad definitions of numerical analysis came about. Leaders of our field at the beginning of the computer era, notably Jim Wilkinson and George Forsythe, discovered that rounding errors sometimes led to surprises, much bigger errors in the final answer than you might expect. They were fascinated by this effect and made it their mission to tell the world. Watch out, computers may fail you! How I wish they had been less effective in their admonitions. For the bigger truth of this subject is that it is all about calculating numbers correctly—generally with amazing speed and by methods that are often not in the least bit obvious. As I mentioned earlier, here is the definition I prefer:

Numerical analysis is the study of algorithms for the problems of continuous mathematics.

You can add qualifications, and in particular, many would distinguish the applied end of the field as "scientific computing," but this definition is the essence of the matter, and the spotlight is on algorithms, not rounding errors. If rounding errors vanished, 90% of numerical analysis would remain.[2]

Years have passed, and my essay has had some influence, as has the improvement of the rounding error situation through the adoption of the IEEE standard for floating point arithmetic. Some of the newer textbooks put less of an emphasis on rounding errors on page 1, and Wikipedia, for example, now gives a definition of numerical analysis that I might have written. The data science/ machine learning revolution is also doing its part to heighten interest in numerical algorithms after decades in which many mathematicians and computer scientists avoided them. Indeed, after years of moving the other way, computer science is growing more numerical these days as it is discovered that numerical methods for optimization, for example, are effective in ever new and unexpected settings. May the progress continue.

6. Discrete and Continuous

The distinction between discrete and continuous mathematics is complex and sometimes hard to pin down, but very important. I think this gulf is as big as the one between pure and applied, and in my case it may be bigger, for I consider myself qualified to

---

[2] This estimate was borne out in the SIAM 100-Dollar, 100-Digit Challenge in 2002, which I'll say more about later, where contestants had to compute ten numbers each to 10-digit accuracy. Exactly one of the ten problems depended on rounding errors in the sense that a successful solution required extended-precision computer arithmetic.

have opinions about pure continuous mathematics, but not about discrete mathematics, pure or applied.

Discrete vs. continuous starts from the difference between counting and measuring, and it is related to the distinction between algebra and analysis. I first heard of this dichotomy in my senior year at Exeter, when Nat and I spent a couple of days visiting our friend Eric Anderson in his freshman dorm to see what Harvard life was like. I was talking with one of Eric's classmates, and he made a remark that seemed deeply wise. The first thing you must decide about yourself as a mathematician, he said, is whether you are an algebraist or an analyst.

I had no idea then which I was, but over the years I have come to know: I am an analyst all the way down. What matter to me are real and complex numbers and the functions associated with them. Algebraic and combinatorial structures are a distant galaxy, one I admire through the telescope. If I worry about the state of mathematics, it's the continuous side I am thinking of. I have no standing to worry about the discrete side.[3]

If you look yet again at the definition of numerical analysis, you'll see that its penultimate word hints at the existence of another field with a dual definition:

Numerical analysis is the study of algorithms for the problems of continuous mathematics.

Plainly the other field will be defined like this:

□ *is the study of algorithms for the problems of discrete mathematics.*

So what is □? The answer is computer science, or at least one of the classical parts of computer science (CS for short). The view

---

[3] That said, one thing does puzzle me. In computational mathematics, although univariate polynomials are ubiquitous, multivariate polynomials are not used very much, as I discussed last year in one of my "Notes of a Numerical Analyst" columns in the *LMS Newsletter*. Yet multivariate polynomials are precisely the subject matter of algebraic geometry, one of the highest-prestige areas of pure mathematics, with ten Fields medals. What's going on?

that the study of algorithms is the heart of computer science was put forward by Don Knuth of Stanford beginning in the 1960s, and Knuth took on a stature in computer science like that of Noam Chomsky in linguistics. He made the analysis of algorithms exciting and intellectually deep, and the discipline of CS was born. Knuth has been a career-long inspiration to me and also a friend, but he is at heart a discrete mathematician, and the continuous side of algorithms got mostly left aside by him and his circle when they were defining computer science in the academic imagination. Alan Turing and John von Neumann, by contrast, computer pioneers of an earlier generation who both died young in the 1950s, were equally at home with the discrete and the continuous.

This leads to an odd situation. Intrinsically, you would think numerical analysis ought to belong to computer science, since it's all about computing. And yet computer scientists are mostly trained in discrete mathematics, not continuous. On the other hand, physicists, chemists, and engineers are trained in continuous mathematics, because that's what they need for their work. And so it happens that numerical analysts employed in Computer Science departments sometimes find they can talk to the faculty in every science or engineering department, except their own.

But lately we are less often to be found in the CS department. When these units were first founded in the 1960s, about half the pioneers were numerical analysts, including Walter Gautschi and John Rice at Purdue, George Forsythe and Gene Golub at Stanford, Fritz Bauer in Munich, Eduard Stiefel and Heinz Rutishauser in Zurich, John Bennett in Sydney, Bill Gear at Illinois, Tom Hull at Toronto, Germund Dahlquist in Stockholm, and Leslie Fox here at Oxford.[4] Likewise, about half the papers in the *Journal of the Association for Computing Machinery* in those days were on numerical topics. This has all changed since

---

[4] Fox was the first Professor of Numerical Analysis at Oxford (1963–83), Bill Morton was the second (1983–97), and I am the third (1997–).

then. The (renamed) *Journal of the ACM* now rarely publishes numerical papers, and at most universities, numerical analysis has moved back to the mathematics department. It happened at Oxford in 2009, on my watch. A year or two in advance of the move, the head of the CS department wrote in a draft summary report for the Research Assessment Exercise that when the Numerical Analysis Group moved from CS to Maths, both departments would be improved! Alas, I made a fuss about this silly insult, and it was deleted from the final document. What a treat it would have been to have left that gem in the historical record.

7. Turing Awards and My Mathematics-CS Oscillation

I wrote one of my index card notes on the question of discrete and continuous in 2019, focusing on the Turing Award, computer science's highest honor. In theory, computer science includes numerical analysis as part of its scope, and a numerical analyst is certainly eligible, in theory, to win the award. How many of the 72 winners have actually been numerical analysts? The answer is three, if you count Richard Hamming (1968). The other two, the "card-carrying numerical analysts" in the group, were Jim Wilkinson (1970) and William ("Velvel") Kahan (1989). Note that these dates are 54, 52, and 33 years ago. The award citations for Wilkinson and Kahan both emphasize their contributions to the problem of rounding errors, that 10% of numerical analysis which people mistake for the whole of the field.

Curiously, although I've only read one work by a Fields medalist, I've read about fifteen by Turing Award winners. On that basis, you might judge that I am a computer scientist. As a second data point, the positions I've held suggest I am a perfect mix of CS and mathematics:

Student at Harvard 1973-77: mathematics
Graduate student at Stanford 1977-82: CS
Postdoc at NYU 1982-84: mathematics and CS

Assistant/associate professor at MIT 1984-91: mathematics
Associate/full professor at Cornell 1991-97: CS
Professor at Oxford 1997- : CS, then mathematics

My books and papers, however, show where my heart really lies. I am a mathematician.

8. Pure and Applied

It's time to say the next word about pure and applied mathematics. I've published a few essays touching on this divide, such as my April 2011 "From the SIAM President" column in *SIAM News*. It's a dichotomy all mathematicians have opinions on, and as with all matters of identity politics, it can be awkward. Here in the Andrew Wiles Building at Oxford, we have adopted euphemisms to avoid uttering the dangerous words: we speak of "North wing" and "South wing" mathematics. Before the building was opened in 2013, there had been talk of assigning offices randomly to encourage interactions between disparate fields, but that idea didn't last long, and the pure people ended up on the north side of the central common room and the applied ones on the south.

As you might expect, most applied mathematicians are concerned with scientific applications. Our South wing includes a big research group known as OCIAM, the Oxford Centre for Industrial and Applied Mathematics, whose origins are in solid and fluid mechanics. We also have the Wolfson Centre for Mathematical Biology and the Mathematical and Computational Finance Group. The work in these research groups starts from applications of mathematics to mechanics, electromagnetics, biology, finance, and other areas. A key word is "modeling," as an applied mathematician of this kind uses mathematics to model the natural, social, or cyber world. The expectation is that the required mathematics will already largely exist, such as the notion of a partial differential equation, and the fundamental scientific laws being wielded will also exist, such as Maxwell's equations for electromagnetism. According to this admittedly simplified way of thinking, the applied mathematics consists in combining these ingredients in nontrivial ways to understand nontrivial phenomena.

Well, this isn't what I do. (There have been two exceptions in my career, involving transition to turbulence in the 1990s and Faraday cages twenty years later.) I am one of those researchers who develop *methods* rather than *models*. The same goes for many numerical analysts, though not all. It is sometimes said that our focus is on "applicable" rather than applied mathematics.

There's nothing wrong or odd about this situation. It is altogether fitting and proper that the mathematical population should include some people interested in models and others in methods, and indeed, in OCIAM and the other applied groups at Oxford you will find faculty members who work on both, such as Jon Chapman, a magician of asymptotic analysis. But the fraction who are primarily oriented toward methods is small.

I remember discussing the models vs. methods question over lunch with Harvey Greenspan one day when I was an Assistant Professor at MIT in the 1980s. Greenspan, a dominant figure in the group and an expert in fluid mechanics, said that numerical analysis wasn't a serious research subject. That feeling used to run deep in some people, the idea apparently being that a good scientist can figure out computational methods on the fly, as needed. Luckily, you won't encounter this opinion so much anymore. But it remains the case that among applied mathematicians, there is sometimes a certain distance from the perspective of the numerical analysts. When Alain Goriely, Professor of Applied Mathematics and Director of OCIAM, showed me a draft of his *Applied Mathematics: A Very Short Introduction*, I saw that the pages on numerical computation gave the impression that the first thing one must know about this subject is that numerical results are often wrong! There's our flair for publicity again. Goriely is a good friend, and when I pointed out to him that this looked odd, he was quick to fix it.

The *usefulness* of numerical analysis, all across science and engineering, is unarguable. At MIT and then again later at Cornell, I taught graduate courses in numerical linear algebra and numerical solution of PDEs that were taken by PhD students from a dozen different departments, for these skills are needed in all scientific areas. At Oxford, there was no concept of inter-departmental courses at the PhD level, but starting in 1999, I decided to introduce such a course anyway, running for two terms with the title "Scientific Computing for DPhil Students." I taught this ten times, and there are more than 500 alumni from the various departments of Oxford's Mathematical, Physical, and Life Sciences Division.

9. Five Mathematical Fields

There are five mathematical fields I need to talk about, starting with four with which my career has had longstanding connections. I regard these subjects as among mankind's lasting achievements. At the same time, in each case, although I have been involved in the area for decades, my relationship with it has been in certain respects odd and unsatisfactory. For most of this time I assumed I was simply at fault for this. More recently, I am not so sure.

1. Approximation theory, beginning with my undergraduate thesis.

2. Complex analysis, beginning with that high school course with David Robbins.

3. Real analysis and partial differential equations. This was the area of my PhD thesis at Stanford and is core territory for numerical analysis.

4. Functional analysis. My work in my "pseudospectra period" 1989–2004 was mainly related to eigenvalues or spectra of matrices and operators, one of the central topics of functional analysis. In my "Chebfun period" since then, I have been concerned with developing continuous analogues of discrete notions from linear algebra, which on the face of it is the very definition of functional analysis.

These four fields cover much of the analysis side of mathematics. How it adds up with the decades! And there is a fifth field I will also speak of, where my involvement is newer but still important to me.

5. Probability and stochastic processes. This became personal with my work leading to Chebfun's randnfun command in 2016.

10. Laboratory Mathematics

Before turning to these areas, I want to say a word about how I do mathematics. My habits began to be formed as I worked on my senior thesis at Harvard under Garrett Birkhoff, "Chebyshev approximation in the complex plane." I had weekly half-hour meetings with Prof. Birkhoff in his office at the back of the mathematics library with an alligator hide on the wall, but the choice of thesis topic was my own. For better or worse, I have done little work in my career guided by more senior figures.

Birkhoff did, however, make a good suggestion when he said I should talk to Phil Davis at Brown University. I duly made an appointment and drove down to Providence, Rhode Island. Prof. Davis was avuncular and smart, and upon hearing my questions, he pulled from his shelf an exactly on-target recent article by one Volker Klotz in the *Journal of Approximation Theory*. It was in German. You don't read German? Davis asked. You should know German! This was a life-changing suggestion for me, for I had always avoided the study of foreign languages, thinking this was for less serious people. I started learning German that summer, and it's been a part of my life ever since, joined later on by French.

Age 21 is a foundational time for anyone, and the senior thesis set the pattern of my career. I was skilled at computer programming, and it was natural that I should explore what could be done computationally with complex Chebyshev approximation. It was the spring of 1977, then, that showed me the way. *I do numerical experiments.* Whatever the topic, I use the computer to guide me. This applies when I am working on algorithms and also when I am working on theoretical problems. For example, I can't imagine investigating the Kreiss Matrix Theorem or Crouzeix's Conjecture without doing computations along the way to keep me on track. I have long marvelled at how most mathematicians prove their theorems without taking advantage of this kind of help. (That's why they need to be so brilliant.)

In that first experience with this modus operandi at age 21, I wrote Fortran codes to plot error curves for complex Chebyshev approximations. The plots led to the thrilling discovery that these curves are nearly circular. Not just circular to a few percent, but to one part in a million or a trillion! The reason I was the first to see this was that I was the first to do the experiments. A year or two later in Zurich and Stanford, I was able to prove a theorem establishing the effect theoretically by building on a result of 1925 that I found in a book in the library, and this led to a new construction that I called Carathéodory-Fejér approximation, which turned out to be related to other developing topics that came to be known as AAK theory and Hankel norm approximation. It also led me to the numerical discovery of a certain number 9.28903 ... which now goes by the name of Halphen's constant (or rather its reciprocal). You can see how a fruitful early research experience like this would be formative.

Similarly, my PhD thesis, on group velocity in finite difference schemes for partial differential equations, had plenty of theorems, but they all sprang from computational experiments that alerted me to certain surprising group velocity effects in numerical discretizations, which I realized could explain the physical basis of a celebrated stability theory of Gustafsson, Kreiss, and Sundström. It was in the lead-up to this thesis that I developed the habit of writing research memos as I work on a topic, three or four pages long, usually presenting numerical experiments with figures. My "Waves" memo series ran from "1.

Wave speeds in finite difference schemes" (25 May 1980) to "49. Slides from oral defense" (6 April 1982). Today, forty years later, I have just finished Rat203 in the current sequence of "ratmemos" devoted to rational functions.

As a third example, I mentioned earlier my pseudospectra period. This sprang from computer "plots of dots," revealing how the eigenvalues of nonnormal matrices, such as nonsymmetric Toeplitz matrices, spread out into a cloud if you perturb the matrices a little bit. Gradually, I came to see that the perturbed eigenvalues don't just tell you about the perturbed problem, but more importantly, they encode information about the behavior of the unperturbed one. This was the beginning of the theory of pseudospectra and my discovery that in many fields involving nonsymmetric matrices and operators, eigenvalues did not have the significance that was generally supposed. Eventually this led to the book *Spectra and Pseudospectra* with Mark Embree. Sometimes it feels as if my whole career has consisted of working out the implications of phenomena revealed by computer plots, an utterly obvious thing to do, but most people don't do it.

The discipline of physics divides familiarly into theoretical and experimental, and everybody understands that both are essential to advancing the field. In mathematics, it could be the same, for many phenomena of interest these days are unobservable except on the computer. One classic example is the famous effect known as chaos, discovered by Lorenz in numerical simulations in 1961, and another is the phenomenon of solitons, special nonlinear waves discovered through numerical simulations beginning with Fermi, Pasta, Ulam, and Tsingou in 1953.[5] On the whole, however, the experimental side of mathematics is weaker than it should be and doesn't always get much respect. Indeed, the phrase "experimental mathematics" sounds dim in my ears, suggesting that this is an activity carried

---

[5] These examples had theoretical and observational precursors, notably Poincaré's study of the N-body problem in the 1880s and John Scott Russell's observation of solitary waves in the Union Ship Canal in Scotland in 1834.

out by people who are unaware of theorems or incapable of appreciating them. That's why I gave this section the heading "laboratory mathematics."

When people contrast theory and experiment in mathematics and physics, they often make a logical error. In mathematics, unlike physics, we have proofs, and that is magnificent. The surprisingly common error is to suppose that mathematics has proofs *instead of* experiments as a source of knowledge, when in fact, it has proofs *in addition to* experiments.

Our laboratory is wonderfully lightweight, for all you need is a computer. You have an idea for an experiment? If you've been in training as long as I have, chances are you can carry it out and get some results within the hour. (The physicists are not so lucky.) Such an experiment, for example, was the first step toward learning that the "random Fibonacci sequences" generated by the equation

$$x_{n+1} = \pm x_n \pm x_{n-1},$$

where each $\pm$ sign comes from a random coin toss, grow at the rate $(1.13198824\ldots)^n$ as $n \to \infty$. My Cornell student Divakar Viswanath, now at the University of Michigan, proved the theorem and wrote the paper, and $1.13198824\ldots$ is called Viswanath's constant.

You might think that all numerical analysts would function as laboratory mathematicians, and of course, some do, but it is curious how many do not. Too often, numerical analysis becomes just another specialty to which a mathematician has decided to apply their talent. Such people may draw remarkably little inspiration from computing, regarding it as a tool for confirming theorems rather than the whole point of the enterprise. When a paper is published in this mode, you'll see 25 pages of equations and theorems establishing the theoretical properties of the method under consideration followed by a few pages of numerical experiments at the end to "verify the results." Very possibly the calculations were done by a graduate student, who may be thanked for carrying out this necessary but onerous task.

11. Approximation Theory: My Early Years

I remember my first encounter with approximation theory. To approximate a function like e^x for values $-1 \leq x \leq 1$, you can use a few terms of the Taylor series, like this,

$$e^x \approx 1 + x + 0.5x^2,$$

and you'll get an accuracy of 0.218. But somewhere along the way, maybe as a sophomore at Harvard, I learned that you can do much better with different polynomial coefficients, like these:

$$e^x \approx 0.989 + 1.130x + 0.554x^2.$$

Now the accuracy is 0.045, five times better. Neat! The theory of such Chebyshev or "minimax" approximations appealed to me, and I decided to write my undergraduate thesis in the area.[6] Approximation theory extends much further than improving this or that calculation by a factor of 5. Ultimately it deals with the foundational question of how we can grasp functions at all.

In my PhD student years, I regarded approximation theory as one-third of my portfolio along with numerical conformal mapping and finite difference methods for partial differential equations. Twelve of my first 18 papers were in this area, most of them with Martin Gutknecht of the ETH in Zurich, with whom I formed a happy collaboration after spending the summer of 1979 there. This was the first of three phases of my career as an approximation theorist:

I: academic polynomial and rational approx. (1977–85),
II: numerical polynomial approximation (2004–17),
III: numerical rational approximation (2017–).

In the first phase, it might have been noted that I came from a nonstandard background, since I was doing my PhD in a numerical analysis group. But I made good contributions, notably in connection with Carathéodory-Fejér approximation, non-

---

[6] Decades later, we can compute the number 0.045 with one line of Chebfun:
```
f = chebfun('exp(x)'); p = minimax(f,2); norm(f-p,inf).
```

uniqueness of best rational approximations, and the behavior of Padé approximations. I went to approximation theory conferences and to workshops at the mathematical research center in Oberwolfach, and I met the leaders of the field. I felt young and inexpert, but I liked learning from those who knew more, and I was publishing a solid series of papers with Gutknecht, and this was a normal good start by a mathematical apprentice.

But my PhD thesis was in a different area, and other, more computational interests called, and in the mid-1980s, I moved on to other things.

12. Approximation Theory: Polynomials and Chebfun

Twenty years later began the stage of my career in which I have really been *using* approximations. In fact, I have probably turned into the leading instance of a mathematician who is applying approximation theory to get things done. I am speaking of the relatively classical, one-variable part of the subject. The multivariate side is being applied by many people these days in connection with deep learning, neural networks, and data science.

The flame was lit by the software system Chebfun. Back in 2001, my DPhil student Zachary Battles had asked for suggestions of research topics for his thesis, and on December 4 of that year, I sent him a message proposing seven possibilities. Zachary was a Rhodes Scholar from Pennsylvania and completely blind, an outstanding computer programmer and an amazingly talented person all around. Number 3 on the list was "A Matlab extension for functions," and this was the topic he chose. Chebfun was born, and by 2006, it was the focus of much of my work.

I will talk later about the conceptual basis of Chebfun. What matters here is that its implementation depends on polynomial approximations. Chebfun carries out numerical computations with functions that are realized on the computer as *chebfuns*, which means polynomials in the form of Chebyshev series of adaptively determined degrees, or concatenations of such objects. At every step, the system approximates functions by these chebfuns to 15 or 16 digits of accuracy. The project was perfectly suited to my interests, and it grew from one student at first to as many as ten students and postdocs during the exciting period 2011-2017. Chebfun is fast and powerful, and it has been very successful, with thousands of users around the world and a new paper that cites it being published every 1.5 days on average, according to Google Scholar.

However, something strange happened. As I started to care about approximating functions, I found myself drifting away from the field of approximation theory. The trouble, to put it bluntly, is that approximation theorists are not very interested in approximating functions. They are interested in pursuing their mathematical ideas to the next logical step. The field has a momentum of its own, quite independent of its apparent raison d'être.

Few fields would seem to have so plain a purpose as approximation theory, as is discussed, for example, by Hilbert's student Paul Kirchberger in his thesis in Göttingen in 1902. (I comment on Kirchberger's views in *Approximation Theory and Approximation Practice.*) But in truth, approximating functions is only a starting point. The larger aim of approximation theory, as with so many academic fields, is to follow certain intellectual trajectories. Here, as always, mathematicians are motivated only in part by the challenge of developing useful tools, and in equal measure by the challenge of *making their results as sharp as possible.* The dual obsessions of mathematicians are to make everything as sharp as possible and to make everything as general as possible.

Let me illustrate the allure of sharpness with the problem of optimal interpolation points. It turns out that if you want to interpolate a function $f(x)$ defined for $-1 \le x \le 1$ by a polynomial $p(x)$, then equally spaced interpolation points are a catastrophically bad choice, whereas Chebyshev points, clustered near $+1$ and -1, are excellent. This phenomenon was explained by Carl Runge shortly before he moved to Göttingen in 1904 as

Germany's first Professor of Applied Mathematics.[7] But are Chebyshev points *optimal*? This is just the kind of question that jumps at a mathematician, so natural and enticing. A century ago it was realized that no, they are not optimal, so the question became, what can be said about optimal interpolation points? Bernstein made a conjecture in 1931 about how they can be characterized, and it was exciting, fifty years later, when Kilgore and de Boor were finally able to prove that the conjecture was true. Students learn these things in approximation theory courses. If you ask an approximation theorist whether Chebyshev points are optimal for interpolation, the chances are they will be aware that no, they are not.

But here's what's crazy. How much better are optimal points than Chebyshev points? Twice as good? 10% better? It turns out that they are *zero percent better*! For low-degree interpolation you may gain a little bit, but this quickly shuts off to no gain at all as the degree rises. So the whole literature of optimal interpolation points is pretty much an academic game. If you ask your approximation theorist about this bucket of rain on the optimal parade, there's a good chance they'll be unaware of it.

Optimal interpolation points illustrate how easily mathematicians are distracted by pretty problems from what one might have imagined they are trying to do. The true but misleading statement "Chebyshev points are not optimal" is an example of what I called an "inverse Yogiism" in an essay I published a few years ago in the *Notices of the American Mathematical Society*. Yogi Berra's

---

[7] Runge was exactly 99 years older than me, as we share the birthday August 30 (along with both Howard Emmons and Mark Embree, mentioned earlier). When Runge's retirement age came up in 1924, the first thought was to appoint another Professor of Applied Mathematics as his successor. However, it was decided that the distinction between pure and applied mathematics was no longer helpful, thanks in part to the successful contributions of Runge himself, and upon a vote of the mathematics faculty, the title reverted to Professor of Mathematics. (See Iris Runge, *Carl Runge und sein wissenschaftliches Werk*, p. 192.) Applied mathematics reemerged at Göttingen with the establishment of the Institute of Numerical and Applied Mathematics in 1967.

celebrated quirky remarks, like "A nickel ain't worth a dime anymore," are statements that are literally contradictory or nonsensical, yet convey a truth. Mathematicians' occupational hazard is the reverse: statements that are literally true, yet miss the point.

At the other extreme from optimal interpolation points, namely, very useful but of little interest to approximation theorists, here is an application of approximation theory with great practical value: calculating the roots of functions. If $f(x)$ is a function defined for $a \le x \le b$, the best method of finding its roots is to approximate it by a polynomial p and then find the roots of p by solving a matrix eigenvalue problem. This method works like magic, and it was proposed by Jack Good in a paper in 1961, then realized in Chebfun four decades later. For example, the commands

```
f = chebfun(@(x) besselj(0,x),[0 1000])
r = roots(f)
```

compute all 318 roots of the Bessel function $J_0(x)$ for $0 \le x \le 1000$ to 15-digit accuracy in 1/40 of a second on my laptop. The 100th root is 313.3742660775.... Are approximation theorists excited by this fruit of their researches? No, they are mostly unaware of it. It has no pedigree connecting it to the classical problems investigated by the founders of the field.

Having gone to approximation theory conferences early in my career, I might have resumed the habit later on as I found myself reconnecting with the subject, but I hardly did this at all. I found the conferences and the work they showcased just too academic, too far removed from practical problems of approximating functions. For a taste of the kind of work I have drifted away from, here are the titles of the first five papers published in the *Journal of Approximation Theory* in the millennium year 2000. (There is a principle that a mathematics paper always has a few names in the title.)

"On the rate of convergence of the generalized Durrmeyer type operators for functions of bounded variation."

"A Korovkin theorem for abstract Lebesgue spaces."

"The maximal Riesz operator of two-dimensional Fourier transforms and Fourier series on $H_p(R \times R)$ and $H_p(T \times T)$."

"Multiple refinable Hermite interpolants."

"Markov-Bernstein type inequalities for multivariate polynomials on sets with cusps."

Now in each case, I can figure out what is going on in the paper with a bit of work, and in fact, it will probably be interesting. But when it comes to computation, this kind of research is not much help. It's no more and no less than perfectly normal academic mathematics.

Thus, as I found myself using approximation theory more and more, I was interested less and less in learning about new research developments. The feeling was mutual. If in 1985 I may have seemed like a young man on the way up, with the potential to become a leader of the field, what I have grown into a third of a century later is quirkier than that. My name is well known, but I am not much of a figure in the approximation theory community, and I am not often invited to speak at the conferences.

Paradoxically, somewhere along the way I seem to have written one of the main textbooks in the field, *Approximation Theory and Approximation Practice ("ATAP")*. I loved working on this project, especially during a sabbatical at TU Berlin hosted by Volker Mehrmann. The book had an unusual gestation. In March 2009, I was invited by Max Jensen to give a seminar at the University of Durham. By that point Chebfun was becoming very good at illustrating concepts of approximation theory, and it occurred to me that for the seminar, it would be interesting to show off some of these capabilities, so I offered the title "Approximation theory and approximation practice." Four years later, the book appeared.

To make *ATAP* strong, I decided to track each idea to its original source and list them all in an annotated bibliography.

This took a good deal of work, for mathematicians have a way of talking at length about the fine points while not mentioning the coarse ones. For example, a function $f(x)$ defined for $-1 \le x \le 1$ has a *Chebyshev series*. Now there are two basic ways to get a degree n polynomial approximation to f: either you truncate the series, or you interpolate f by a polynomial in $n + 1$ Chebyshev points. (This is a version of the "coefficients" vs. "values" distinction we will meet again later.) Existing textbooks of approximation theory—an excellent and appealing group, by the way, mostly written about fifty years ago—do not point out that there are these two possibilities. They do not tell the reader how simple the relationship between the two is, involving the so-called "aliasing formula," nor show that they are equally effective in approximating functions (to within a factor of 2), nor present theorems for the two cases in parallel. I had to figure all this out for myself, even if in the end many of the key facts turned out to have been worked out a century ago and published in papers like "Über einen Satz des Herrn Serge Bernstein" (Marcel Riesz, 1916).

13. Approximation Theory: Rational Functions

And then, six years ago, I began to get involved with rational functions as never before. Rational functions are ratios of polynomials, $r(x) = p(x)/q(x)$. I could tell a number of stories, but I'll concentrate on the oddest, the story of Donald Newman's startling theorem of 1964.

It had been known forever that whereas polynomials are good at approximating smooth functions, they are terrible with non-smooth ones. For example, suppose you want to approximate the absolute value function $f(x) = |x|$ for $-1 \le x \le 1$ with an error no greater than 0.001. You'll need a polynomial of degree $n = 282$, and as the degree increases further toward ∞, the error decreases only in proportion to $1/n$. This is awful, and it means that polynomials are useless for doing anything practical with non-smooth functions, except at very low accuracy.

Then along came Donald Newman's four-page paper in the *Michigan Mathematics Journal.* Newman proved that if you take a *rational* approximation to the absolute value function $|x|$ for $-1 \leq x \leq 1$ of degree n, which means a ratio p/q where p and q are of degree n, then the convergence can be as fast as "root-exponential," that is, errors decreasing at a rate $\exp(-C\sqrt{n})$ for some constant $C > 0$. The speedup is spectacular. Degree $n = 8$ is now enough for accuracy 0.001, and degree 26 for accuracy 0.000001. What a difference!

But now comes something astonishing. Approximation theory is about approximating functions, right? And Newman had just published a result showing that rational functions are spectacularly good at doing that, right? So immediately in 1964, with great excitement, people must have started to apply rational functions for all kinds of computations, right?

Not at all. Newman's result had no impact whatever on numerical computation. And as for the approximation theorists:

They didn't apply Newman's theorem. They sharpened it.

As I say, this is what mathematicians do. None of us showed any interest in developing algorithms exploiting root-exponential convergence. Instead, the attention was on the theoretical notion of exactly optimal "best" approximations, which are hard to calculate and therefore somehow extra-interesting. A number of people started investigating these best approximations of $|x|$ for $-1 \leq x \leq 1$. Vyacheslavov in 1974 proved that the sharp constant is $C = \pi$. Varga, Ruttan, and Carpenter in 1993 found more precisely, through numerical experiments, that the error behaves asymptotically like $8\exp(-\pi\sqrt{n})$, a calculation for which they had to work in 200-digit arithmetic precision. Herbert Stahl proved this result rigorously a few months afterward. Later, in 2003, Stahl generalized the result to the equally academic problem of best rational approximation of $|x|^{\alpha}$, where α is a positive number.

Forty years had now elapsed since Newman. Approximation theorists had sharpened and generalized his discovery. But none of us had applied it to do anything useful! And yet it was potentially extremely useful, for whenever you solve a PDE in a domain with corners—and in practice most domains have corners—there will generally be singularities of the same nature as those in Newman's problem (branch points). So his discovery has a close link to problems of scientific interest, but we had not noted this. Nor had we investigated methods for finding practical approximations with root-exponential accuracy, as opposed to the exactly optimal best approximations that are so fascinating but so hard to compute.

I speak of "we," for I was as distracted as the rest. Rational approximation had always been an interest of mine, and I'd met Donald Newman at a meeting in 1985 in the English village of Shrivenham, where he kindly praised me for finding a rhyme for that name in a limerick. But for thirty years, despite knowing and admiring Newman's theorem, it did not occur to me to try to use it. So I too, like everybody else, pretty much forgot what was supposedly the purpose of approximation theory.

Somehow in 2016, my perspective changed and I began to consider rational functions properly for computation. First Yuji Nakatsukasa, Olivier Sète, and I devised the "AAA algorithm" for rational approximation in ordinary 16-digit arithmetic, which has opened many doors. Then my student Abi Gopal and I adapted Newman's discovery to create what we call "lightning solvers" for PDE problems in domains with corners. The idea is to solve the Laplace, biharmonic, and Helmholtz equations to many digits of accuracy via rational approximations with exponentially clustered poles near each corner, just as in Newman's approximations of $|x|$, and the name "lightning" speaks to the mathematical link to the way lightning strikes trees and buildings at sharp points. Further developments have been contributed by Peter Baddoo, Stefano Costa, Yuji Nakatsukasa, and André Weideman.

All this could have been done forty years ago. I could have shown it to Donald Newman!

14. Graduate Student at Stanford: Gene Golub, Serra House

After Harvard I wanted to do a PhD in numerical analysis. My plan was to go to Berkeley, but one day in the spring of my senior year, Gene Golub telephoned long-distance from California and told me Stanford was better. It thrilled me to get a call from a famous professor, and he was friendly and enthusiastic, so Stanford is where I went. If Velvel Kahan had called, I would probably have gone to Berkeley.

Enrolling at Stanford involved a switch to a department of computer science, since that was the home of numerical analysis. Later, Gene told me I'd been the top PhD applicant in CS that year, but I had no sense of such things at the time, and indeed, I had not heard of linked lists or other basic topics of CS 101, nor had I heard of the department's greatest star, Don Knuth. Students these days seem to manage their careers more scientifically than I ever did, juggling the data carefully to optimize their options. I juggled my feelings and my ambitions, but not much data. In those days we didn't have rankings of universities from 1 to 500 to look up on the internet.[8]

Gene Golub was an extraordinary personality, off the scale in the degree of his attention to people. He was a bachelor, and his students and colleagues in numerical analysis were his life. He was big and friendly and told everyone to "Call me Gene" as soon as he met them. It seemed there were weekly parties at his house

---

[8] We had the internet, though, which was then called the ARPANET, and I even had an address on it in 1978, among the first few thousand: I was CSD.TREFETHEN@SU-SCORE. Only a small group of universities and national labs were connected, but these included Stanford and Harvard, so I was able to exchange emails with my girlfriend back home (an alumna of Math 55—so much for testosterone). I don't remember what we called the messages, but it wasn't a word as short as "email."

at 576 Constanzo Avenue, and in later years he and I kept in touch, frequently having dinner together at a restaurant with some of his other numerical friends. Indeed, whether I was in Oxford or Sydney or Paris, Gene had a way of turning up for an extended visit. He often had a gift in hand, like the latest Robert Caro biography. Golub was a regular part of my life until his death in 2007, and a big influence on me even if we worked in different areas. I was unusual in Gene's circle in never writing a paper with him, and he joked, not without reason, that I was a Harvard snob.

It was because of Gene Golub that I became involved with SIAM, and in fact, four Stanford numerical analysts have served as Presidents of SIAM and several more as Trustees. It was because of him that I attended the first of the quadrennial ICIAM congresses (International Congress of Industrial and Applied Mathematics), a series he played a role in establishing. That was in Paris in 1987, and 35 years later, I am one of only two or three people who have attended all nine ICIAMs.[9] Next year I plan to extend the streak to ten.

Stanford Computer Science in 1977. Somehow I had landed in the heart of Silicon Valley in its early days, when TeX and WIMP interfaces and SUN Microsystems and Silicon Graphics, Inc. were being created. I took my place in a legendary group of graduate students alongside Marsha Berger, Petter Bjørstad, Dan Boley, Ken Bube, Tony Chan, Bill Coughran, Bill Gropp, Eric Grosse, Mike Heath, Randy LeVeque, Franklin Luk, Stephen Nash, and Michael Overton—also affiliated students Jonathan Goodman, Nick Gould, and Jorge Nocedal. Many of these people are now famous in our field. The whole group of us, together with Gene's many visitors, had our desks in a spacious former family home called Serra House, with a persimmon tree in the courtyard.

---

[9] Sir Michael Atiyah, one of four people to date to win both the Fields medal and the Abel Prize, gave a plenary talk at this first ICIAM congress in which he remarked that applied mathematics feeds off the crumbs dropped from the table of pure mathematics. That comment got some attention.

The visiting professors in my first year at Stanford included Germund Dahlquist and Jim Wilkinson, towering figures of the numerical analysis of the 20th century.[10] And I haven't mentioned my thesis supervisor, the generous and appealing Joe Oliger, or the other faculty members in the group, Jack Herriot and later my good friend Rob Schreiber. It was Gene who brought these people together, all united by a love of numerical analysis. Later at Oxford, this was my vision of how an academic research group ought to be.

The importance of Gene Golub in the history of numerical analysis is related to a historical transition and to a clash of terminology. Classically, the roots of linear algebra are in algebra. For example, eigenvalues of matrices are traditionally regarded as algebraic quantities, having elegant invariance properties going back to the 19th century. However, linear algebra is also a subject of analysis, where the corresponding objects are *singular values*, with the *singular value decomposition (SVD)* being the tool you need if you want to measure the size of a matrix or its inverse. The SVD was barely known when Golub started advocating it in the 1960s, but now it is used throughout the computational sciences. This quantitative, analytical side of the subject is the one that matters to us numerical analysts, so for the purposes of this essay, "linear algebra" belongs to analysis, not algebra. Gene preferred to avoid the A-word entirely and talk of "matrix computations."

---

[10] I took two courses from Wilkinson, who was tied with the complex analyst Max Schiffer as my favorite lecturer at Stanford. Wilkinson, with his impish English smile, was a laboratory mathematician at heart. He spoke rivetingly about his experiments on the Pilot Ace computer beginning in 1949, trying to figure out why the errors behaved as they did and eventually coming up with the explanation he called backward error analysis.

15. Complex Analysis and Peter Henrici

Continuous mathematics is about functions, and functions live in the complex plane. This is the plane of complex numbers $z = x + iy$, where x is the real part of z, y is the imaginary part, and i is defined by that magic formula $i^2 = -1$. "Living in the complex plane" is a way of saying that, although you can examine a function like $f(x) = \sin(x)/(1 + x^2)$ for real numbers x, you won't fully understand its properties until you view it as $f(z) = \sin(z)/(1 + z^2)$. Mathematicians have understood this since Cauchy, Weierstrass, and Riemann in the 19th century. For example, the ideas of *integrals* and *infinite series* come fully into their own in the complex plane.

As I've mentioned, I was lucky to be exposed to this subject so early. I liked it very much, so it was natural that I should choose to write my undergraduate thesis on complex Chebyshev approximation. And then, with the help of a word about me from his friend Birkhoff, I got to know the Swiss mathematician Peter Henrici (pronounced hen-REE-tsee), who was writing his great three-volume work *Applied and Computational Complex Analysis*. Henrici spent the fall of 1978 visiting Stanford, and at his suggestion, I set to work on the topic of numerical Schwarz-Christoffel conformal mapping. I spent every weekend that fall at the SLAC computer center developing my algorithm, reporting my progress to Henrici in the weekdays. The project resulted in one of the first robust numerical methods and computer codes for conformal mapping, which later evolved into the SC Toolbox for Matlab by Toby Driscoll, now at the University of Delaware. It also led to my first publication, which appeared in the inaugural issue of Golub's new journal, then called the *SIAM Journal on Scientific and Statistical Computing*. This paper has some nice computer-generated pictures of conformal maps in it, and all my later papers also have computer figures, with just one or two exceptions. This is typical for many numerical analysts.

I published the preprint of the Schwarz-Christoffel paper in March 1979 as CS-TR-1979-710 in Stanford's computer science technical report series, and I believe this was the third research report ever formatted in TeX, which Don Knuth had just created. The first was Knuth's TeX manual STAN-CS-78-675,[11] in November 1978, and the second was CS-TR-1979-703, by Bengt Aspvall. For decades now, TeX (or rather its variant LaTeX) has been the universal typesetting system for mathematicians, physicists, and computer scientists.

Henrici and I hit it off. He was a European professor in his fifties used to dominating the room, and I was a 23-year-old kid, but we had in common, first of all, a love of numerical mathematics. When programmable pocket calculators appeared, he was the first to write a numerical analysis book for them (*Computational Analysis with the HP-25 Pocket Calculator*, 1977), just as when Matlab came out, I would be the first to publish a research paper with a Matlab program in it ("Matlab programs for CF approximation," 1985). Equally important, we shared a love of words, writing, and typing. Later, when Henrici was back in Zurich, we typed dozens of mathematical letters back and forth to each other on our IBM Selectric typewriters. (He had an older model, without the erase key.) It was at Henrici's invitation that I visited the ETH in 1979, and for a couple of weeks that summer, when he was off in the mountains on Swiss military duty, he lent me his office in the Hauptgebäude with its great windows overlooking the Polyterrasse and the city and lake of Zurich. There was a shelf by the desk with a foot of books he had written, and I remember paying quite a bit of attention to that shelf.

To my lifelong regret, Henrici died early, at age 63 in 1987. I didn't realize when he was alive how much I meant to him, for it's easier for an older man to see himself in a younger one than the other way around. Henrici liked my American freshness and my confidence in doing mathematics on the computer. On hearing how I had computed SC maps by combining Golub and

---

[11] Donald E. Knuth, Tau Epsilon Chi, a system for technical text, November 1978.

Welsch's code for Gauss-Jacobi quadrature with Powell's code for quasi-Newton solution of nonlinear systems, he exclaimed with his big laugh that this was a "symphony of computation." The graduate students in his group at the ETH, who were older than me and in fact in two cases married with three children, probably wondered why this American upstart was getting so much attention. Henrici insisted I must call him Peter, though his students didn't get that offer until the completion of their PhDs. I remember him confiding in me one day that he felt the younger people around him somehow didn't like words very much. He thought they talked too slowly, and he liked my verbal ease.

I wish Henrici had lived to a normal lifespan. I've been close to several senior figures over the years, notably Gene Golub as mentioned and also Gil Strang and Cleve Moler, but Henrici and I had something special. It never got a proper chance to grow.

When I brought my draft of the Schwarz-Christoffel paper to his office toward the end of his time at Stanford, formatted in TeX of course, I had put both our names on it. "Oh, I don't need another publication," Henrici said generously. So I removed his name and became the sole author; but now I wish I had insisted.

16. Complex Analysis: CAvid and CMFT

I've been working with complex variables ever since. Sometimes this is for applications that are obviously complex, like conformal mapping or Padé approximation, but equally often it is for problems that on the face of it involve just real numbers, yet where you need the complex context to get it really right. For example, in my papers on Clenshaw-Curtis quadrature, on functions in the d-dimensional hypercube, and on solving Laplace's equation in a polygon, the algorithms, theorems, and proofs all rely on complex variables.

In numerical analysis we like algorithms that converge quickly as you increase the number n of steps or parameters, and in the best cases, this means they converge at an exponential rate when the functions they are applied to are smooth enough.

Smooth enough means *analytic*, the condition of having convergent Taylor series at each point. This is where the subject of complex variables comes in, for if a function is analytic, it can be extended from the real line into the complex plane, and exponential convergence can then often be proved by analysing a complex contour integral.

For example, the most famous method for calculating a real integral $\int_a^b f(x)\,dx$ is known as Gauss quadrature, which involves sampling f at n points between a and b and then adding up the samples multiplied by certain weight factors. Here is the striking result from the 19th century:

> *Theorem. If f is analytic, Gauss quadrature converges exponentially.*

This theorem tells us that Gauss quadrature has the most basic good property we could ask for.

But do you know something odd about this fundamental theorem? It appears in almost no textbooks!—although almost every numerical analysis text teaches the reader about Gauss quadrature.[12] One reason may be that the notion of analyticity is considered too advanced, even though it has been the basis of mathematicians' understanding of functions for 200 years. Another may be that the authors of the textbooks don't know the theorem, since they learned the subject from previous textbooks. In fact, numerical analysts are often quite weak in complex variables. Maybe it wasn't like that in the 1950s and 60s, but one can only do so much, and the vital subject of numerical linear algebra has expanded so greatly since then as to take up a lot of the oxygen. Today's numerical analysts speak fluent Matlab, and they know how to precondition a Krylov subspace iteration, but

---

[12] Folkmar Bornemann has pointed out that the theorem can be found as Corollary 5.3.5 in *Numerical Methods in Scientific Computing I*, by Dahlquist and Björck (2008).

most of them haven't touched a contour integral since their student days.

What about numerical analysts like me who use complex variables all the time? No doubt there is a community of experts we can call upon for advice?

This is where it gets painful. Yes, there are hundreds of specialists in complex analysis around the world. They know more than I do about many things, and sometimes I am able to call upon their expertise. Yet usually, when I have tried, I have not gotten very far. Our languages and value systems are too different.

Complex analysis is not regarded as one of the hot areas of mathematics these days, and a number of these specialists probably feel isolated in their departments. A typical paper in this area has less impact than one of mine, at least as measured by citations, so you might imagine that the experts would be eager to be in conversation with me. That's not how human nature works, though. They are trained in one way of thinking and I in another, and to each of us, the other's concerns seem not so important.

If you look at the research topics investigated by complex analysts, incidentally, you'll see plenty of words that appear computational. Complex analysts "estimate" and "compute" all the time, as do many other mathematicians. But these estimates and computations are conceptual ones. Computation in the sense of actually working with numbers is a fringe activity.

In my three-month sabbatical at the University of Geneva in 2014, I had an office next door to the Fields medalist Stas Smirnov, one of the most exciting complex analysts of all, but I only managed to talk with him once. I was working on the mathematics of the Faraday cage effect, which had inexplicably never been sorted out since Faraday's original discovery in 1836; even Richard Feynman in his *Lectures* gets it wrong. This topic is intimately tied to complex variables, but I was unable to catch Smirnov's interest. If the same problem had been brought in the door by a colleague from his own area, his engagement might have been different, but as a numerical analyst, I was not in the set of people he was predisposed to pay attention to.

During the Covid pandemic, a worldwide "CAvid" lecture series has been superbly organised by Rod Halburd, the "Complex Analysis video seminars." Unfortunately, after an initial burst of enthusiasm, I have missed many of these talks, which have titles like "Loewner-Kufarev energy and foliations by Weil-Petersson quasicircles." When I see a title like that, I suspect the speaker and I may have little in common. So I fall into the habit of not paying attention, and undoubtedly, as a result, I miss some things that would be interesting.

I gave a CAvid talk of my own and decided to begin with some personal remarks about these difficulties. Here is how I put it:

> I want to say a word about our field, the field of complex analysis, and me. As Rod says, I'm a numerical analyst, but pretty much everything I've done, or let's say two-thirds of the things I've done over the years, have been rooted in complex variables. That is my central playground and maybe my main advantage as a numerical analyst, because many numerical analysts aren't so good at complex analysis. But the personal thing I wanted to say is a somewhat sad one, which is that I have very little connection with this community, the community of more theoretical complex analysts. Most of you on this call I don't know; I haven't met three-quarters of you. I don't read your papers, with a few exceptions, and probably you don't read mine. It's amazing how separate the practical computational world is from the theoretical world. That can't be good. I don't have a solution to offer, but it can't be good, and in particular, for me personally, it's meant I haven't benefited from experts as much over the years as I should have. So I work on a project; it's very much in the complex plane; I know there must be experts out there I should be asking. I usually don't know who they are and I usually don't manage to ask them. What a waste.

Another impressive enterprise in complex analysis is CMFT, which stands for Computational Methods and Function Theory and is both a series of quadrennial international conferences since

1989 and a high-quality journal since 2001. (Function theory is another name for complex analysis.) From the start, the announced vision has been to blend the numerical and the theoretical:

> CMFT is an international mathematics journal which publishes carefully selected original research papers in complex analysis (in a broad sense), and on applications or computational methods related to complex analysis.

I was excited when this vision was announced, and I happily accepted the invitation to be one of the inaugural editors of the journal. Unfortunately, the "C" in CMFT has proved to be silent. People like me have ended up playing little role, and the articles published in the journal only occasionally have any true engagement with applications or computational methods. The editorial board now has 53 members, only one of whom, Lothar Reichel, is known as a numerical analyst. I think the fault for this situation lies on both sides. The purer complex analysts would like to be closer to computation, but they don't know what steps to take, and the computational ones like me would like to be closer to theoretical developments, but we don't know which ones to tune in to. As I write, the ninth CMFT conference has just taken place in virtual mode. None of the invited speakers were numerical people, and apart from my own, few of the contributed talks showed evidence of actual computation. I believe the live demo in my talk was the only one at the conference.

Rainer Kress of the University of Göttingen has pointed out to me another example of the "silent C" phenomenon. One of the oldest mathematics journals is the *Journal für die reine und angewandte Mathematik*, founded in 1826. Although the name means "Journal for Pure and Applied Mathematics," this journal is actually pure.

But to return to complex variables. Fortunately, there is a small group of good friends who share my joy in computing in the complex plane, including Peter Baddoo, Stefano Costa, Tom DeLillo, Toby Driscoll, Bengt Fornberg, Nick Hale, Cécile Piret,

Alex Townsend, Elias Wegert, André Weideman, and Heather Wilber. Take a look at the spectacular "Complex Beauties" calendars by Wegert and his colleagues and you'll see what I mean.

17. Postdoc at NYU with Peter Lax: Pure and Applied Again

Oh, mathematics is beautiful. Real analysis, the study of functions of a real variable, has such powerful theorems! Continuity, compactness, Fourier transforms,… the elegance and importance of these topics is deeply satisfying. And real analysis leads to the language of the laws of nature, partial differential equations or PDEs. When Maxwell discovered how light waves work, it was because of a PDE. When Einstein predicted gravitational waves, it was because of a PDE. Chemistry is built on Schrödinger's equation, fluid mechanics on the Navier-Stokes equations, and civil engineering on the equations of elasticity.

My PhD thesis at Stanford was in this area, specifically, the numerical solution of hyperbolic PDEs. For a postdoctoral next step, the glamorous place to go was therefore the Courant Institute of Mathematical Sciences at New York University (NYU), in Greenwich Village, New York. Unlike most mathematics departments, the Institute focused on just one area, namely real analysis, PDEs, and their numerical analysis, and in its constellation of stars, the most brilliant was Peter Lax. Lax was a wunderkind from Hungary who had worked in the Manhattan Project as a teenager. Now 56, he was in his prime, and I found myself at Courant for two years with him as my NSF postdoctoral supervisor. I also taught a course each year as an Adjunct Assistant Professor.[13]

If every mathematician were like Peter Lax, there might be no need for this essay. His brilliance and charm set the tone for the Institute, where he was a social as well as an intellectual magnet.

---

[13] It was at this time, back again on the same coast as my father Lloyd MacGregor Trefethen, that I started going by my middle name, Nick.

At lunch we would gather with him in the lounge on the 13th floor, and the conversation would be one of substance. Sometimes he would first lead a party to Dean & DeLuca's in Soho to pick up the right ingredients, for of course, he had gourmet tastes and wanted to welcome others into his world. I remember his twinkly eye and curly hair and curiosity about all subjects. In his central European way, he seemed to know everything about music and literature too.

I didn't work with Lax, for he had half a dozen postdocs on the roster and my tastes were more computational than his. But his influence on me was still great, and when I say that if everyone were like him things might be different, I mean something specific. Lax's mathematical mind encompassed both pure and applied. His publications had a pure style, centering on theorems proved with technical perfection, but he knew and appreciated the applied things too. He had enormous impact on numerical analysis through celebrated theorems that encapsulated just the right point in each area, like the Lax equivalence theorem, which I had studied as a graduate student.

There is a widely held view I mentioned earlier, that mathematics is one, that the difference between pure and applied is illusory. This is nonsense, and in my experience it is generally an opinion held by pure mathematicians, who often fancy themselves applied, or fancy they could easily be applied if they chose. You know that joke, what's the difference between an entomologist and an etymologist? An etymologist knows the difference. I think there's a principle like this in mathematics. What's the difference between a pure and an applied mathematician? An applied mathematician knows there's a difference.[14] Our mission, the mission of this essay, is to encourage better communication between the many parts of mathematics without pretending they are all the same.

---

[14] In his 1981 essay with the notorious title "Applied mathematics is bad mathematics," Paul Halmos makes the same distinction—but in the other direction!

Lax was brilliant and wide-ranging enough to pull it off. If we were all like him, mathematics might indeed be one, but most of us are not so remarkable.

Pure mathematics has its eyes on history, preferring to work on ideas that will still be important in 100 years, which for good reasons may often be very abstract and general. For the people at the top, this orientation may be realistic, and indeed, it is one of the triumphs of the human spirit that we can create mathematics that lasts for centuries. For more average researchers, however, this model is not always a good one, and contributes to mathematicians having so much difficulty in understanding each other's work.[15]

18. Real Analysis and PDEs: Regularity

Having given quite a few pages to two of my five fields, I am at risk of going on at frightening length with the remaining three. To avoid that, I'll confine myself to one observation for each. For real analysis and PDEs, my theme is smoothness, or as mathematicians call it, "regularity."

How smooth is a curve? The basic idea for answering such questions is derivatives. If a function is continuous, that's not much smoothness at all, but if you can differentiate it, i.e., take a derivative, that is good. If you can differentiate it twice, that is better. And so the basic measure of smoothness is the number of derivatives you can take, and there are standard notions such as $C^k([a, b])$, the set of functions defined for $a \leq x \leq b$ whose k^{th} derivatives exist and are continuous. Such ideas apply not just to functions of a single variable, i.e., curves, but also to functions of several variables, namely surfaces.

---

[15] I've served on committees to award fellowships for pure mathematicians. To explain why a candidate is deserving, a referee will begin by making an attempt to describe the substance of their achievements, but this is hard. Pretty soon the letter moves away from substance and resorts to asserting how brilliant the candidate is. Every discipline judges people in part by their brilliance, but no other takes it as far as mathematics.

The degree of smoothness of a function need not just be an integer like 0, 1, or 2. We can talk about a function having "half a derivative" of smoothness by appealing to a notion known as Hölder continuity. There's also another technology going by the name of *Sobolev spaces*. With Sobolev spaces you can consider in a systematic way the sets of all functions that can be fractionally differentiated, say, $1/2$ or $\sqrt{2}$ or π times, which are represented with the notations $H^{1/2}$, $H^{\sqrt{2}}$, and H^{π}. The mathematics is elegant and, of course, completely rigorous. If you want further refinements, there are Besov spaces $B_{p,q}^{s}$ and Triebel-Lizorkin spaces $F_{p,q}^{s}$. I remember at one point in my postdoctoral stage thinking it was very important that I should understand the details of Besov spaces.

Now, why do we go to the trouble of this delicate analysis? At the outset, the mathematics forces us to. For example, suppose you have a function f and you want to work with its Fourier series, a decomposition into an infinite collection of sines and cosines. Naturally you ask, does the series converge to f? It turns out that for convergence to be assured, it's not enough for f to be continuous, but it's more than enough for f to be differentiable. So of course mathematicians want to work out a sharp criterion for exactly how much smoothness is needed. It turns out that any amount of differentiability is more than enough, like half a derivative or one-millionth of a derivative. So more refined analyses come into play of functions that don't have any differentiability at all, yet are still a little smoother than just continuous. It's a long and technical story, addictive in its complexities, and many mathematicians have contributed to it. Barry Simon's textbook on harmonic analysis, which is what this subject is called, runs to 759 pages, and it is not too extreme an oversimplification to say that its central project is to develop theorems relating different measures of smoothness of functions to different convergence properties of their Fourier series and transforms.

So regularity theory starts from natural questions, but it has grown into a monster, consuming everything in sight. Mathematicians are made fun of for worrying about the existence of a solution without caring about how to find it, but an equally good caricature would be that they worry about regularity. *How smooth is this object?* Though scientists and engineers hardly care, this question in a thousand forms dominates real analysis and PDE theory. It gets much more attention than questions you might have thought were more fundamental, like, How good a model of a scientific problem is this equation? How can we find solutions? What do the solutions look like? What phenomena do they reveal?

As ever, it's a case of mathematicians' attraction to the challenges of sharpness, generality, and technical difficulty. I've just looked up the papers published so far this year in the Springer journal *Partial Differential Equations and Applications* and the NYU journal *Communications on Pure and Applied Mathematics*. Five of the 17 mention regularity *in their titles*.

A theme of this essay has been that we often see the pure mathematicians doing one thing and the numerical analysts another. But regularity theory for PDEs is an exception, for here, the numerical analysts have followed the theorists. I am speaking especially of the dominant technology for solving PDEs known as the Finite Element Method. In the finite elements numerical analysis literature, you will rarely see a problem even stated, let alone investigated, in any terms except Sobolev spaces. If it's a fluid mechanics problem, the velocity may be assumed to belong to H^1, the pressure to H^0, and the pressure gradient to H^{-1} (a space of functions with "minus 1 derivative"). The finite element discretization and its convergence theory will be tuned to these spaces. Everything fits together perfectly, with all the pieces interlocking in an elegant fashion.

It's impressive, but how distant from the functions that arise in applications! Let me explain. Years ago, in the days of Euler and Lagrange, the default assumption was that a function would be given by a formula, which means essentially that it's analytic.

I think of this as the 18th century notion of a function. As mathematics developed, the default assumption swung to the other extreme, that a function is merely continuous, and I think of this as the 20th-century notion of a function. Having, say, one derivative or half a derivative is a minor variation on this assumption.

But what functions usually look like in practice is something else again:

"18th c. function"   "20th c. function"   what functions usually look like
analytic   continuous   piecewise analytic

They are analytic, not merely continuous, except at isolated points (or curves or surfaces, in higher dimensions) where they have jumps or other singularities. Just think of a rectangle, the simplest domain where a PDE problem of scientific interest is likely to be posed. At the corners, the solution will probably have singularities. Along the sides, it will probably be analytic. Functions of this kind have almost no place in real analysis or PDE theory. In fact, even the big book by Grisvard on analysis of PDEs in domains with corners, precisely the setting where you might think a different notion of functions might be called for, begins with 80 pages of Sobolev spaces. And so our mathematical analysis, and our finite element algorithms in their standard forms, fail to recognize or exploit the perfect smoothness that so many functions have almost everywhere in their domains.

Do you know how difficult it is to construct a function that is merely continuous, or has merely one or two derivatives, all along its length as permitted by the Sobolev theory rather than at isolated points? Until a celebrated example of Weierstrass was published in 1872, it was not even known that this was possible. Nowadays the preferred method is to make use of the mathematical idealization of Brownian motion, in which infinitesimal random pulses nudge the curve up or down at every point along the way, and that's how the 20th-century sketch above was made. Of course, there are applications where this is just what is needed, but these are the exceptions. Yet every time a PDE theorist or a numerical analyst investigates a problem in the setting of Sobolev spaces, they are implicitly working with this pessimistic model of functions and probably settling for algorithms with correspondingly low convergence rates.

19. Cleve Moler and Matlab

En route to functional analysis and Chebfun, I must mention Cleve Moler and Matlab.[16]

I first met Cleve Moler when I was a graduate student and he visited Stanford, where his loud and friendly voice reverberated around Serra House. Moler is the antithesis of a European, and as a transatlantic soul, I love both Europeans and their antitheses. A room with Moler in it is a no-nonsense zone. He has no interest in showing you how your problem is connected with the theory of pseudodifferential operators. He just wants to get things done computationally, and nobody has done it better. Moler is about the same age as Knuth, and while Knuth was writing his great books on the analysis of discrete algorithms, Moler was creating the modern era of numerical software. He was an author of both of the foundational software packages of the 1970s, EISPACK and LINPACK, and he also published two influential software-based numerical analysis textbooks. And then, in around 1977 in the Computer Science department at the University of New Mexico, he invented Matlab, which changed the world.

Matlab started as an interface to portions of EISPACK and LINPACK. Instead of requiring programmers to invoke Fortran subroutines through elaborate calling sequences, the idea was to let them compute interactively at the terminal, typing commands like `eig(A)` to find the eigenvalues of a matrix or `A\b` to solve a

---

[16] The correct orthography is MATLAB, but I don't like that, so I write Matlab instead.

linear system of equations.[17]   All the right algorithms would be invoked in all the right places, without the user needing to know the details. At first Matlab was regarded by many people as a toy, good for the classroom but not for "real" computing. But before long it was a programming language as well as an interactive system, and once Moler, John Gilbert, and Rob Schreiber gave it sparse matrix capabilities in 1992, Matlab became a tool for serious numerical computing of the desktop scale, as opposed to the supercomputer scale needed for, say, weather prediction or analysis of chemical molecules.

Moler visited Stanford again on sabbatical in the winter quarter of 1978-79 and taught CS238b, which met MWF 12:00. I was in the class along with Marsha Berger and Randy LeVeque as he explained matrix eigenvalue algorithms with demonstrations in Matlab. At the time, I was working on approximation theory and conformal mapping and beginning to think about PDEs, and I'm not sure Matlab made much of an impression on me. But it certainly impressed the engineers in the class, and a couple of years later, when I was in my Assistant Professor office at MIT, I remember Cleve coming in to introduce a young man to me. "This is Jack Little," he said. "He's starting a company to sell Matlab!"[18]

At that stage, at MIT, I was the only numerical analyst on the applied math faculty and I was teaching the numerical linear algebra course, with a Sun-1 workstation to play with in the basement. I had money to spend, having been named as a Presidential Young Investigator, and when the chance came along

---

[17] For me $eig(A)$ epitomizes the successful contribution of numerical analysis to our technological world. Physicists, chemists, engineers, and mathematicians know that computing eigenvalues of matrices is a solved problem. Simply invoke $eig(A)$, or its equivalent in whatever language you are using, and you tap into the work of generations of numerical analysts. The algorithm involved, the QR algorithm, is completely reliable, utterly nonobvious, and amazingly fast. On my laptop, for a 1000×1000 matrix A, $eig(A)$ computes all 1000 eigenvalues in half a second.

[18] The IBM PC had been introduced in August, 1981.

to buy Matlab from the new company, I placed an order for ten copies for $500. (One of the licences was employed by my student Alan Edelman, now a professor at MIT, in his classic work on condition numbers of random matrices.) Only a decade later did MathWorks inform me that I had been their very first customer. They gave me a plaque, now on display in my office: "First order for MATLAB, Professor Nick Trcfcthen, February 7, 1985." Another one is on the wall at the MathWorks headquarters in Massachusetts.

I said Matlab changed the world, certainly the world of desktop computing by numerical analysts, applied mathematicians, and engineers. In particular, it changed my research life. If I had no strong reaction to it as a graduate student, that situation transformed when I was a junior faculty member with a workstation, which by now had moved upstairs to my office. I found that Matlab fitted my research and teaching style perfectly. The numerical experiments that had started in Fortran with my undergraduate thesis at Harvard now had a more natural platform, and it became a part of my mathematical life I have never ceased to rely upon. It's around 37 years now, let's say 14,000 days, and I would estimate I have used Matlab on 12,000 of them.

20. Ten Digits

Digits of accuracy have always fascinated me, for they are the stamp that you have solved your problem. Of course, some problems can be solved exactly in the sense of an analytic formula, but these are exceptional. Most of the time there is no formula, and one must compute. This starts with problems as simple as finding a number x that is equal to its own cosine, i.e., $x = \cos(x)$. (Solution: $x = 0.7390851332....$)

Two years after landing as the Professor of Numerical Analysis at Oxford and in charge of the Numerical Analysis Group, I started a tradition that we kept up for fifteen years. Each October, five or six new students would arrive to begin DPhil (PhD) studies with me or one of the other faculty members. As an American used to graduate students broadening their knowledge by taking courses for a year or two, I didn't like the British system of these 21-year-olds getting right down to full-time research on problems that were often all too academic. So I decided to require them to give some time in their first term to what we called the Problem Solving Squad. Each week for six weeks, I handed out a problem, usually stated in just one or two sentences, whose solution was a single number to be computed numerically. There were no hints. The students' challenge, working in pairs, was to compute each number to as many digits of accuracy as they could. Here are some of the problems.

A particle starts at the top vertex of a triangular array with 30 *points on each side, and then takes* 60 *random steps. What is the probability that it ends up in the bottom row?*

What is $\sum_{n=2}^{\infty} \sin(n) / \log(n)$?

Three regular tetrahedra each have volume 1. *What's the volume of the smallest sphere you can fit them inside?*

What is $\int_0^1 \sin^2(\tan(\tan(\pi x)))dx$?

A needle of length 1 *rests on the surface defined by the height function $h(x) = 0.1x^2 + 0.1\sin(6x) + 0.03\sin(12x)$. What is the lowest possible height of the center of the needle?*

What is the smallest value of $\varepsilon > 0$ for which the equation $\varepsilon u'' + u - u^3 = 0$ with $u(\pm 1) = 0$ has exactly five solutions?

What is $\sum n^{-1}$, where n is restricted to those positive integers whose decimal representation does not contain the substring 42?

At what time t_∞ does the solution of the equation $u_t = \Delta u + e^u$ on a 3×3 square with zero boundary and initial data blow up to ∞ ?

If $f(x,y) = \exp(-(y + x^3)^2)$ and $g(x,y) = \frac{1}{32}y^2 + e^{\sin y}$, what is the area of the region of the x-y plane in which $f > g$?

Two adjacent solid unit cubes, each with mass 1, *attract each other gravitationally according to Newton's Law with constant $G = 1$. What is the force of attraction between them?*

Non-mathematicians may not recognize how unusual, even strange, some of these problems are. They have no scientific motivation, and no ordinary mathematical motivation either. What they have is *algorithmic* motivation. They aim to test whether the students can figure out enough about the structure of the problem to really "nail" it. The Squad participants were encouraged to look up any and all sources of information and to talk to friends and faculty members. We had some long weeks of effort and some very satisfying successes. Now and again I made a mistake in cooking up a problem the night before—for example, one problem turned out to have the answer ∞, which I failed to spot in advance—but fortunately, most of them made good sense. There were also a couple of problems where an exact solution was unexpectedly found. The most surprising example of this was the "two cubes" problem above, for which Bengt Fornberg of the University of Colorado later derived an insanely complicated exact formula consisting of a sum of fourteen terms along the lines of $35 \log(1 + \sqrt{5})$ and $22 \tan^{-1}(2\sqrt{6})$. To test the correctness of his solution, of course, we compared it against the numerically computed result 0.9259812605 .... This story is told in a chapter of the 2011 book *An Invitation to Mathematics*, edited by Dierk Schleicher and Malte Lackmann, and also in one of my *LMS Newsletter* columns in 2020. Wider context of the problem can be found in work of Michael Trott of Wolfram Research, Inc. and Folkmar Bornemann of TU Munich.

After the Problem Squad had been running a few years, in 2002, I decided to organize a digit-hunting event for people outside Oxford. I selected ten problems and posted them in *SIAM News* as the "SIAM 100-Dollar, 100-Digit Challenge." Contestants, who could work in teams of up to six people, had to try to solve each problem to ten digits of accuracy, and their score would be their total number of digits. The Challenge attracted a good deal of attention, and twenty teams got perfect scores of 100 points. (All 20 won $100, thanks to an anonymous donor who was later revealed as William Browning.) The story is compellingly told, including mathematical details of the problems going far beyond anything I had had in mind, in *The SIAM 100-Digit Challenge: A Study in High-Accuracy Numerical Computing* by Folkmar Bornemann, Dirk Laurie, Stan Wagon, and Jörg Waldvogel (SIAM, 2004). In the course of writing their book, Bornemann et al. managed to solve nine of the problems to 10,000-digit accuracy; the tenth remains stuck at 273 digits. Further developments were described by Bornemann in 2016 in "The SIAM 100-digit challenge: a decade later."

As the Challenge became known, people started to associate me with the project of computing numbers to ten-digit accuracy, and I realized this was a philosophy I believed in. I think of 3 digits as "engineering accuracy," what you might hope for in a problem with complex geometry and physics, whereas 10 digits is "scientific accuracy," a good target when the problem is more idealized. Three digits are usually plenty for an application, but they are nowhere near enough if you are building a computational foundation for further work. There is also an algorithmic divide between 3 and 10 ten digits. While many algorithms may solve a problem to low accuracy, such as the randomized simulation known as Monte Carlo analysis, you usually won't be able to get 10 digits unless you have the mathematics more fully under control. Another consideration is that, in physics, whereas many quantities are known to 5–10 digits of accuracy, like the speed of light or Planck's constant, not many are known much beyond that: so 10 digits is a reasonable proxy for exactness. Finally, there is the convenient feature that 10 digits is well below 16 digits, the level of rounding errors, so it can usually be achieved by computation in standard floating point arithmetic.

I wrote an essay on this philosophy of numerical computing called "Ten digit algorithms," in which I defined a TDA by three conditions:

Ten digits, five seconds, and just one page.

A Ten Digit Algorithm should fit on one page of code in your computer language, and it should compute the answer to 10-digit accuracy in no more than five seconds. A lot of thinking went into this definition, and in particular, the five-second condition requires the computation to complete on a human time scale, so that a good researcher will be unable to resist adjusting parameters, exploring, confirming. (So much error results from people settling for experiments that take minutes or hours to run!) My essay was issued as an Oxford Numerical Analysis Group report in 2005, and it ends with a bullet list and three sentences of exhortation:

> Ten digit algorithms can
>
> - Improve our publications
> - Speed up program development
> - Make our numerical methods faster
> - Make our scientific results more reliable
> - Facilitate comparisons of ideas and results
> - Add focus to the classroom
> - Add zest to our field.
>
> The challenge of designing these codes raises our standards and raises our expectations. It's good for the academics, and it opens the doors wider to non-academics. And it's fun!

Apart from the technical report, the TDA essay has not been published, and in fact, it has the distinction of having been rejected by arXiv. I tried twice to post it there, and both times, the repository responded by saying that they regretted that the piece was not substantial enough. Evidently, to be rejected by arXiv, it is not necessary to find a flaw in Einstein's theory of relativity or discover new properties of the number 666.

21. Functional Analysis: Chebfun's Blank Slate

Thanks to Wilkinson, Golub, Moler, and other mathematicians, and to EISPACK, LINPACK, Matlab and other computational tools, numerical linear algebra is a booming business. Its methods are widely known and taught to new generations around the world, partly from my own textbook *Numerical Linear Algebra* co-authored with David Bau when I was on the faculty at Cornell, before moving to Oxford. One explanation of this success is that, to a degree unpredicted by von Neumann and the other pioneers, computational science comes down to linear algebra. Your scientific problem may be formulated in nonlinear partial differential equations, for weather prediction perhaps, but you'll probably reduce it in two steps to get answers on the computer:

Linearization: nonlinear → linear

Discretization: analysis → algebra

It's the second step, discretization, that is the terrain of Chebfun. As the schema suggests, often when we are dealing with discrete vectors and matrices, they are only discrete because we have made them so to fit on the computer. We would prefer to deal with their continuous counterparts, functions and linear operators, and the vision of Chebfun was to make this possible. Just as we compute with numbers like e and $\sqrt{7}$ without thinking about how they are approximated in 64-bit floating-point arithmetic, we would like to be able to compute with functions like $\sin(x)$ and e^x without thinking about how they are discretized with respect to x.

What opened the door to realizing this vision was the introduction into Matlab of Object-Oriented Programming. A central idea of OOP is *overloading*, where you take an operation and give it a new meaning without changing the syntax. Matlab's syntax already encapsulated matrix algorithms developed over generations. How about retaining that framework but overloading the operations, and introducing appropriate new algorithms, so that our programs would work with functions and operators instead of vectors and matrices? We would be doing continuous linear algebra at the keyboard, which is so often what the user wanted in the first place.

Chebfun unfolded in a manner somewhat like that of Matlab itself. At first, our expectation was that the experiment would be interesting but perhaps not so useful, since surely the execution would be slow. We were surprised to find how fast it actually was. Before long we had overloaded 100 Matlab commands to their continuous counterparts, including mainstays of numerical linear algebra like the singular value decomposition. Everything was mathematically and algorithmically new and had to be figured out. Big steps towards practicality came when we allowed functions to be piecewise rather than just globally smooth and to depend on two or three variables rather than just one. The key people in these developments were my DPhil students Ricardo Pachón and Alex Townsend and my postdocs Rodrigo Platte and Behnam Hashemi. Another big step came when we overloaded the Matlab backslash operator for matrix systems of linear equations, $x = A \backslash b$, to solve ordinary differential equations (ODEs), $u = L \backslash f$. After an initial spark from Folkmar Bornemann of TU Munich, this differential equations side of Chebfun was built by Toby Driscoll and later Ásgeir Birkisson and Nick Hale. (Hale, now at Stellenbosch University, directed the big release of Chebfun Version 5 in 2014 and has written more Chebfun code than anyone else.) Without it ever having been planned, Chebfun has emerged as the most convenient software tool available for solving ODEs. Further Chebfun contributions were made by DPhil students Anthony Austin, Nicolas Boullé, Abi Gopal, Hrothgar (mononymic), Mohsin Javed, Hadrien Montanelli, and Mark Richardson, by postdocs Jared Aurentz, Silviu Filip, Pedro Gonnet, Stefan Güttel, and Kuan Xu, by my close faculty colleague Yuji Nakatsukasa, and by sabbatical visitor Grady Wright from Boise State University. Heather Wilber also extended Chebfun to work with functions in a disk when she was just a Masters student at Boise State.

For fifteen years, this continuous mode of numerical computing has been my world. It's not just Matlab I use daily, but Matlab with Chebfun. I don't know how long Chebfun itself will last, but the idea of overloading vector operations to functions is here to stay.

This brings me to functional analysis, one of the major areas of mathematics since its initiation by Fredholm, Hilbert, and Schmidt at the turn of the 20th century and the focus of another of Oxford's mathematical research groups. Functional analysis could be defined as

The study of continuous analogues of linear algebra.

This isn't how you'll usually see it put, but it's the essence of the matter. This is the field of mathematics in which we do linear algebra on functions instead of discrete vectors.

Do you see a possible link to Chebfun?

Strangely, there has been very little link at all. For an explanation of this situation, we must note that for a mathematician, the word "continuous" implies the transition from finite-dimensional to infinite-dimensional spaces, and anything infinite is a rich source of technical challenges. Indeed, the need to treat infinities rigorously, which arose for example with Cantor's theory of sets and Lebesgue's theory of integration, is perhaps the biggest single reason why mathematics had to become more technical in the past 150 years. These challenges, in all their richness, are the wellspring of the field of functional analysis. For example, the notion of the eigenvalues of a matrix in linear algebra, if you want to make it rigorous for continua, becomes the more advanced notion of the spectrum of a linear operator, which may be divided into the point spectrum, the continuous spectrum, and the residual spectrum. Other matrix concepts like nullspaces and ranges likewise acquire new complications. Big theorems come into play, like the Hahn-Banach theorem and the uniform boundedness principle, which as a rule express results too trivial to mention in the linear algebra case. The monumental work

Linear Operators by Dunford and Schwartz fills three volumes, 2592 pages.

In all those pages, and all the achievements of functional analysis, there were probably things that could have been of use to us in developing Chebfun; but how were we to find them? Generations of mathematical results addressing genuine mathematical problems proved strangely distant from our project of creating a continuous analogue of linear algebra on the computer. As I write, this term's talks in our department's functional analysis seminar series have just been announced, with titles like "Applications of subfactor and categorical techniques to C*-algebras" and "Schatten class Hankel operators on the Segal-Bargmann space and the Berger-Coburn phenomenon." It's another world. Once again, I stop paying attention, and, no doubt, occasionally there are things I miss.

For Chebfun, we started from a blank slate and built it all ourselves.

22. Stochastic analysis: Values vs. Coefficients

My fifth and final area is different from the others. With those, I can claim some expertise bought with the years, but in this case I was mostly a newcomer when I got involved in 2016.

The mathematics of probability has been with us for a long time, and in the 20th century it took great strides, but still it remained a specialized subject among mathematicians. For example, when I was a postdoc at the Courant Institute in the 1980s, although a probabilist or two were on the payroll (Henry McKean was a star, and Raghu Varadhan would later win the Abel Prize), everyone knew that the main subject was PDEs. But somehow in more recent years, probability has moved to center stage. You see this in mathematics departments around the world, and starting in 2006, a regular fraction of Fields medals have been awarded for advances related to probability, not to mention that Abel Prize to Varadhan and another one recently to Furstenberg. At NYU, among 72 professors currently shown at the mathematics web site, 14 list PDEs among their interests and 14 list probability and/or stochastics. Half the research talks now seem to have a probabilistic flavor, including those in the numerical analysis/scientific computing seminar. I just checked and found that the next talk in this series has about as probabilistic a title as you could imagine: "Leveraging concepts from stochastic simulation and machine learning for efficient Bayesian inference."

I am about 50% cynical about the rise of probability. Partly I think it is important and exciting, and that machine learning, for example, indisputably requires probabilistic foundations. It also seems that for computational problems of a sufficiently large scale, the best algorithms almost always make use of randomness. The other half of me wonders if the trend is driven more by the hunger for ever bigger problems than by genuine scientific need. It sometimes seems as if our computers have grown too powerful for the old problems to be challenging, so we make them harder by upgrading constants to random variables. Recently I marked a student paper that explained with the blitheness of youth that whereas in the past, scientists modeled phenomena with differential equations, now it is recognized that these need to be replaced by stochastic differential equations.

Yet plenty of stochastic analysis is well justified, and it can certainly be interesting. Indeed, there's a fascination in random phenomena that I think reflects something deep in our psychology. If you simulate a random process, you'll find it seems to have a personality like a living creature, and you'll want to investigate further.

A few years ago, the Chebfun project faced a challenge. Like every computing system, Matlab has a command that produces random numbers. If you type `randn(1000,1)`, you get a vector of 1000 of them. What's the continuous analogue? What should the output be from the Chebfun command `randn`, or as we ended up calling it, `randnfun`?

The sketch above shows our answer, for it is a plot of the output from the Chebfun commands `cumsum(randnfun(0.001))`. The "cumsum" part of this instruction specifies an indefinite integral, the continuous analogue of a cumulative sum, and `randnfun(0.001)` specifies an approximation to white noise. Mathematicians call such a curve a *Brownian path*, which is a random walk in the limit of infinitely many steps of infinitely small size. Albert Einstein, Jean Perrin, and other physicists understood the essential features of Brownian paths soon after 1900, including the intriguing property that they are continuous but nowhere differentiable. Norbert Wiener in the 1920s began the process of making the theory rigorous. Stochastic analysis is still under development, and for example, Martin Hairer's Fields medal in 2014 was awarded for his rigorous treatment of nonlinear stochastic partial differential equations.

Now to describe any function, we always have the two options of "space and Fourier space," or "space and dual space," or as the Chebfun team likes to say it, "values and coefficients." You can represent $f(x)$ by its *values* at each point x, or as an infinite series, in which case you are working with its *coefficients*. This duality goes back to Joseph Fourier in Egypt with Napoleon more than 200 years ago (or perhaps to Alexis Clairaut in the 18th century), and it has proven to be one of the most fruitful ideas in all of mathematics, even if there are technical challenges involved, as I mentioned earlier in connection with the 759 pages of Barry Simon's book.

For the `randnfun` command, we realized that we would have to use the "coefficients" approach, since a chebfun must be smooth, being represented by polynomials. The specification we settled on was for `randnfun` to output a chebfun corresponding to a finite Fourier series with random coefficients, with the length of the series specified by a parameter. In the call to `randnfun` above, the parameter is 0.001, meaning that there are on the order of 1000 terms in the series.

As always, there is the alternative "values" approach to Brownian paths, in which we evaluate $f(x)$ at individual points, and the parameter is how many points we evaluate. The two approaches are mathematically equivalent, as was established by Wiener himself, even if only one of them lends itself to Chebfun implementation.

Stochastic analysis comprises not just random functions, but also random differential equations. These are called SDEs, short for stochastic differential equations, and here there are the same two approaches. You can represent solutions by coefficients, and Chebfun does this, providing (unexpectedly) one of the world's simplest tools for exploring SDE phenomena, as described in our *SIAM Review* paper of 2017 and in chapter 12 of our book *Exploring ODEs*. Or you can represent them by values, and this was the method that was developed first, by the Japanese mathematician Kiyosi Itô in the 1940s. Later, in the 1960s, Eugene Wong and Moshe Zakai proved that the values and coefficients approaches to SDEs are equivalent.

Intellectually, this has been a fascinating journey for me, finding that Chebfun could do something useful in the area of random functions and encountering along the way some powerful mathematics. What a bit of luck to learn about stochastic differential equations by discovering we had created a good tool to solve them!

Sociologically, the experience has been less positive. As always, to learn what I needed, my instinct was to ask questions of everybody, both acquaintances reachable by email and the local experts in the common room. I am afraid I found these conversations difficult. I quickly discovered, stochastic analysts do not like the "coefficients" approach. They do not use it, they do not teach it to their students, and they do not put it in their textbooks, even though it is simpler than the Itô and Stratonovich calculi required by the "values" formulation, not to mention the associated Euler-Maruyama and Milstein numerical methods. Indeed, it was only slowly that I learned that both formulations were in the literature and had been proven to be equivalent.

Everyone who has taken a calculus course knows the idea of integrals, but only those who have studied more advanced mathematics know the subject of *measure theory*. The "values" formulation of stochastic analysis has technical complications closely allied to those of measure theory, yet this is the formulation that mathematicians insist must be taught from the beginning. I have been unable to determine why their feeling is so strong on this point. Imagine if we told students they had to learn measure theory before they could talk about integrals![19]

As I say, I found it difficult to discuss these matters with the stochastic analysts. There I was in the common room, having understood eventually that one can define Brownian paths and SDEs by either values or coefficients and eager to learn the pros and cons of the two approaches—eager to learn why stochastic analysts, biologists, and financial mathematicians (but not physicists) work with the first rather than the second. Instead of enjoying stimulating conversations about a subject that is absolutely fascinating, I felt I was being told to go away and study Gaussian processes, or regularity structures, or rough paths.

Do you see what is so puzzling here? We all know how trying it can be to have to explain something to a person who lacks the necessary background and asks foolish questions. But is there a discipline other than mathematics where we would expect to see this dynamic play out between senior professors in the same department?

---

[19] This is the final sentence of our *SIAM Review* paper.

23. Mathematics Today

It would be hard to argue that I am anything but a mathematician. And yet I have described a career in which, as the decades have passed, I have become established in this profession while feeling I am drifting away from it.

In my early years I took it for granted that the more mainstream mathematicians, the leaders in each specialized field, understood what was important in their areas. It troubled me, therefore, to notice that my own work wasn't building on theirs. I would investigate a problem and make a good contribution, often a genuine discovery, without ever mastering or in the end even attempting to master the results of the nonnumerical experts in the area. Indeed, I joked occasionally that

Numerical analysis is the study of the mathematics of the previous century.[20]

Despite the joke, privately I interpreted the situation as a deficiency on my part. I knew I was doing good work, but I supposed it would be even better if I had the strength of character to absorb the papers of Adamjan, Arov, and Krein in support of my Carathéodory-Fejér approximation, to immerse myself in the theories of the great Louis Nirenberg while I was working on PDEs at the Courant Institute, or to digest Dunford and Schwartz when I was writing the book on pseudospectra. Year after year, I had the sense I was falling short.

One may spot a possible flaw in that reasoning. If ignoring the masters were truly an error, then I would have found not infrequently in my career that my contributions later turned out to have been anticipated, or invalidated, by the work of others. This has not happened. Everything I've done has remained valid and original, some things more important than others, of course, but almost never mistaken or redundant. Indeed, it is clear that if I

---

[20] This kind of floating reference appeals to the mathematical sense of humour. A leading textbook on fractals takes it further with the dedication, "To my current wife."

had done more of *that*, I would have done less of *this*. We are all finite, and for better or worse, tying myself more keenly to the mathematics of my day would have made me a different researcher. I would have been more of a pure mathematician and less of a numerical analyst.

Reflecting on this phenomenon of worthwhile contributions from such different communities, I find it hard not to conclude that the division of labor reported at the beginning of this essay, that pure mathematicians develop the concepts and numerical ones develop the algorithms, is oversimplified. At many points in my career I have found that the established concepts were not on target: that Atiyah's crumbs travel, or at least ought to travel, in both directions. For example, I've mentioned how eigenvalues proved not to have the significance generally supposed for nonsymmetric matrices and operators, whereas pseudospectra come closer. A second example: slowly over the years I came to realize that, although I have written a number of papers and even edited a book on numerical conformal mapping, this is not actually a good method for its most famous application, solving PDEs in complicated domains. A third example: in working with approximations of functions in squares and cubes and hypercubes, I found that the standard notion of the degree of a multivariate polynomial, the "total degree," is not appropriate; instead one should use the "Euclidean degree," a term proposed by Jared Aurentz. A fourth example: discovering that, despite a century of literature presenting Gauss-Hermite quadrature as the optimal method for the integration of functions on the infinite real line $-\infty < x < \infty$, it is in fact much less efficient than other, simpler methods like the trapezoidal rule. This last experience exemplified a common feature of inverse Yogiisms, that sometimes you can see quickly that the established formulation is askew, but then it may still take quite a bit of work to pin down exactly where the problem lies.

So perhaps I have not wasted my time as a mathematician, but this does not resolve the puzzle. What in the world is going on with mathematics if careful attention to the works of the leaders of approximation theory, complex analysis, real analysis/PDEs, functional analysis, and stochastic analysis need not be on the path to making contributions in these fields?

I don't fully know the answer, but here is perhaps a start. A discipline is defined in part by its *subject matter*, and for continuous mathematics, though some would no doubt say this is oversimplified, I think the subject matter is

Numbers, functions, and equations.

But it is also defined by its *methodology*, and for any kind of mathematics, two quite distinct methodologies might be said to be

(A) *Theorems and proofs*

and

(B) *Algorithms and computations.*

I believe my experience has shown that to an extent one would hardly have thought possible, (A) and (B) can operate independently and still successfully, and they have been doing so for a long time, producing valid advances on both sides that have strangely little to do with one another. Of course, there is a certain amount of communication in both directions. But some mathematicians are carrying forward mathematics mainly in the setting of (A), while others are making equally genuine contributions in the context of (B). Those of us on the (B) side do not abjure theorems and proofs, and indeed we often create our own, as I have done many times for many topics. But for the most part we ignore the theorems and proofs that define the cutting edge of today's mathematics as it is generally understood—Fields medal mathematics, if you like.

What an odd situation! Researchers on both ends of the mathematical spectrum manage to be productive, I am glad to say, and at every point in-between; but still, it is hard not to wish this degree of separation could be diminished.

As a dual citizen of the USA and the UK, I see pure and applied mathematics in terms of an analogy. Nobody would say that the American and British societies are, or should be, identical. But they are certainly related, and surely both will be stronger if there is good communication between them. The current state of pure and applied mathematics feels rather as if America and Britain communicated only by occasional sailboats passing back and forth.

Some people believe that theoretical physics is in a state of crisis brought on by the dominance of string theories unsubstantiated by experiments. I think the situation in today's mathematics is similar, with a detachment of a bewilderingly large fraction of our community from phenomena coupled with an extreme attachment to abstraction and technique. The word crisis, however, is probably excessive in both cases. Physics and mathematics are among the great ongoing achievements of humanity, and both remain alive and strong. Concerning the one that I know, mathematics, I only wish to suggest that its current state is not all it could be, and that it is a challenge for the next generations to bring the field closer together.

Acknowledgments

This essay was written in the early months of 2022 with the encouragement of SIAM's Executive Editor Elizabeth Greenspan. Many friends and colleagues commented on drafts along the way, correcting mistakes, offering new observations, and improving the tone. My heartfelt thanks go to Peter Baddoo, Folkmar Bornemann, John Butcher, Rob Corless, Tom DeLillo, Toby Driscoll, Patrick Farrell, Nathaniel Foote, Abi Gopal, Alain Goriely, Nick Gould, Nick Hale, Mike Heath, Des Higham, Nick Higham, Don Knuth, Rainer Kress, Randy LeVeque, Philip Maini, Kate McLoughlin, Cleve Moler, David Mumford, Yuji Nakatsukasa, Michael Overton, Gerlind Plonka, Lothar Reichel, Michael Saunders, Robert Schaback, Rob Schreiber, Gilbert Strang, Steve Strogatz, Endre Süli, Alex Townsend, Jacob Trefethen, Divakar Viswanath, Elias Wegert, and André Weideman. Several of these people offered detailed, page-by-page suggestions, leading to many interesting discussions. Nat Foote was particularly supportive at a crucial early stage, and Kate McLoughlin is the best editor I know.

Index